KNITTING
WITH
COTTON

KNITTING WITH COTTON

MELINDA COSS and DEBBY ROBINSON

SIDGWICK & JACKSON
LONDON

First published in Great Britain in 1988 by Sidgwick & Jackson Limited

ISBN 0-283-99613-7

Produced by Justin Knowles Publishing Group
9 Colleton Crescent, Exeter, Devon, EX2 4BY

Design: Vic Giolitto

Illustrations: Kate Simunek
Photography: Adrian Peacock
Hair and make-up: Jalle, Ellie Wakamatsu and Sarah Matheson
Art direction and styling: Brenda Knight

Typeset by AB Typesetting Ltd, Exeter, Devon
Colour reproduction by J. Film Process Co. Ltd, Thailand
Printed in Italy by New Interlitho

for Sidgwick & Jackson Limited
1 Tavistock Chambers, Bloomsbury Way,
London WC1A 2SG

CONTENTS

INTRODUCTION

Cotton is produced from the fluffy white fibres that surround the seed of the *Gossypium* plant. These seed-pods, or "bolls", are ready for harvesting when they split open to reveal the cotton in its raw state. The processing and spinning may then commence, and the resulting yarns may have a matt finish (sometimes still referred to as "dish-cloth" cotton, as old-style dish-cloths were, indeed, made out of minimally processed cotton yarn) or a silky finish, which is the result of mercerization.

During the process of mercerization, the raw yarn is first passed through a concentrated caustic soda solution. It is then restretched and rinsed to avoid further shrinkage and facilitates dyeing. Mercerization gives a smoother lustre to the yarn, and this may be accentuated by twisting the individual plies of the yarn, which are spun together to produce the finished item. A very tight twist results in "cable" types of yarn, which are much heavier than those with a looser spin because of the density of the strand.

Although we are all familiar with cotton fabrics – cotton accounts for some two-thirds of the world's total textile consumption – cotton yarns for hand-knitting purposes have become widely available only relatively recently. Because of this, many knitters are nervous of using cotton yarns, suspecting that they will behave quite differently from woollen ones. In fact, cotton yarns are interchangeable with any others of a comparable thickness – e.g., a 4-ply wool may be swopped for a 4-ply cotton. As with so many aspects of knitting, however, successful knitting in whatever yarn is totally dependent on achieving the exact tension that is stipulated in the pattern. To produce the same tension, it may be necessary to change the size of the needles used, since cotton does not have the elasticity found in wool yarns.

This lack of elasticity means that cotton requires even greater care in working an even, regular stitch, because any small faults, such as a stretched or split stitch, will be far more visible in cotton work. Yarns with natural elasticity have a "pull in" effect, which causes the stitches to sort themselves out after working, so disguising many irregularities. Since cotton knitting will not do this, you must pay greater attention to the work and place a complete ban on bad habits such as putting your work down part-way through a row or pushing your needles through the knitting when it is not being worked – both guaranteed ways of producing unwanted holes.

When using a thick cotton yarn, the weight of the work may make it slightly more difficult to handle, and a very heavy garment will tend to drop slightly after completion; so make allowance for this when choosing a style.

When working intarsia or fairisle fabrics with cotton, it is even more important to follow the guidelines given on pages 13–14 to avoid gaps forming in the work when you change colour and overtight floats when you carry a colour.

Cotton's lack of elasticity may also affect the overall look of a stitch, especially in lacy patterns, which will appear looser because any holes or strands produced will appear larger and more obvious. Other stitches may be enhanced by the use of cotton, however, especially those that create textured fabrics, because they will have a greater crispness than the same stitch worked in a softer yarn.

7

The major problems caused by cotton's lack of elasticity are baggy welts and ribs. Welts and ribs can become very stretched, as can the cast-on edges, which frequently become "frilly". A firm cast-on edge, as recommended on pages 14–15, must be worked to begin with, and if you tend to have difficulty working a neat rib, use a smaller needle than the size quoted. Looseness may be remedied after knitting by the addition of shirring elastic to the back of the ribs. Use a tapestry needle so that the elastic does not show on the right side of the work. Take care not to pull the elastic too tight or it will snap when any strain is taken. A colourless "knitting elastic" is now available on the market, which is designed to be knitted in as a second strand with the working yarn. This sounds like the answer to all problems, but be careful: it does alter the appearance of a rib and can result in some prize-winning tangles.

Cotton is an incredibly robust fabric, and most cotton knits, unlike woollens, may be machine washed. Great care must be taken when it comes to drying, however. Give cottons only a very short spin and then dry flat, having first pressed the garment into shape. If you are washing by hand, always support the wet weight of the garment, squeezing it against the side of the bowl rather than holding it up. Never hang up a cotton knit to dry but roll it in a towel to remove any excess moisture before laying it flat. Cotton garments are dry-cleanable but, because they are often light summer colours, you should use a reputable dry cleaner whose cleaning fluid is changed regularly, ensuring that the colours will not be left looking dingy.

ABBREVIATIONS

alt	alternate(ly)		p	purl
beg	begin(ning)		p-b	purl into back of st
cb	cable back		psso	pass slipped stitch over
	(*see* Techniques, pages 16–17)		rep	repeat
cf	cable front		rev st st	reverse st st
	(*see* Techniques, pages 16–17)		RH	right hand
cm	centimetre(s)		RS	right side
cont	continue/continuing		sl	slip
dec	decrease/decreasing		ssk	slip 1, knit 1, pass slip stitch over
inc	increase/increasing		st(s)	stitch(es)
k	knit		st st	stocking stitch
k-b	knit into back of st		tbl	through back of loop(s)
LH	left hand		tog	together
MB	make bobble (*see* Techniques, page 15)		WS	wrong side
			yb	yarn back
ml	make one – i.e., inc 1 st by working from the st below the next st to be worked		yfwd	yarn forward
			yo	yarn over needle
			yrn	yarn round needle

TECHNIQUES

READING THE GRAPHS

Throughout the book explanatory graphs show the colour designs charted out, with stitch symbols added where necessary. Each square represents one stitch across, i.e., horizontally, and one row up, i.e., vertically. The graphs should be used in conjunction with the written instructions, which will tell you where and when to incorporate them. Any colours required or symbols used will be explained in the pattern. Always assume that you are working in stocking stitch unless otherwise instructed.

If you are not experienced in the use of graphs, remember that when you look at the flat page you are simply looking at a graphic representation of the right side of your piece of work, i.e, the smooth side of stocking stitch. For this reason, wherever possible, the graphs begin with a right side (RS) row so that you can see exactly what is going on as you knit. Knit rows are worked from right to left and purl rows from left to right.

TENSION

Knitting is simply a series of connecting loops, the construction of which is totally under the knitter's control. Tension or gauge is the term used to describe the actual stitch size – its width regulating the stitch tension measurement and its depth regulating the row tension measurement. Obtaining a specific tension is not a magical skill denied to all those but the initiated. It is a technicality, the controlling factor being the size of needles used by the knitter.

Since all knitting instructions are drafted to size using mathematical calculations relating to one tension and one tension only, that tension must be achieved before you start the work or you will have no control whatsoever over the size of the finished garment. *This is the most important rule of knitting*.

At the beginning of every pattern, a tension measurement will be given, using a specific stitch and needle size – e.g., "using 5mm needles and measured over st st, 18 sts and 24 rows = 10cm square". You must work a tension sample using exactly the same stitch and needle size as quoted. Cast on the number of stitches plus at least two extra because edge stitches, which do not give an accurate measurement, will not be counted. When it is complete, lay the tension sample or "swatch" on a flat surface and, taking great care not to squash or stretch it, measure the tension, using a ruler and pins as shown.

If there are too few stitches, your tension is too loose. Use needles that are one size smaller to work another swatch. If there are too many stitches, your tension is too tight. Use needles that are one size larger to work another swatch.

Even if you have to change needle sizes several times, *keep working swatches until you get it right*. You save no time by skipping this stage of the work, because, if you do so, you risk having to undo an entire garment that has worked out to the wrong size. You may feel that a slight difference is negligible, but a tension measurement that is only a fraction of a stitch out per centimetre will result in inaccurate sizing because each fraction will have been multiplied by the number of centimetres across the work.

If you have had to change your needle size to achieve the correct tension for the main stitch and if other parts of the garment are worked on different sized needles, you must adjust these needles by the same ratio. For example, if you are using needles that are one size smaller than are quoted for stocking stitch, use needles that are one size smaller than are quoted for the ribs.

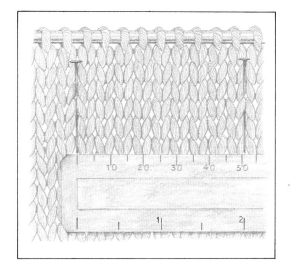

Use a ruler and pins to measure the tension of a sample piece of knitting.

The wrong side (WS) of the work, showing stranding at the correct tension.

The wrong side of weaving, showing the up and down path of the carried yarn.

We have intentionally omitted detailed reference to row tension because many people worry over this unnecessarily, changing their needle size even though they have achieved the correct stitch tension. Although important, row tension does vary considerably from yarn to yarn and knitter to knitter. If your stitch tension is absolutely accurate, your row tension will be only slightly out. Nevertheless, keep an eye on the work, especially when you are working something like a sleeve, which has been calculated in rows rather than centimetres, and compare it with the measurement diagram in case it becomes noticeably longer or shorter.

FAIRISLE

The technique of colour knitting called "fairisle" is often confused with the traditional style of colour knitting that originated in the Fair Isles and took its name from those islands. Knitting instructions that call for the fairisle method do not necessarily produce a small-motifed repetitive pattern similar to that sported by the Prince of Wales in the Twenties – far from it, as can be seen from some of the patterns in this book.

The method referred to as fairisle knitting is when two colours are used across a row, with the one not in use being carried at the back of the work until it is next required. This is normally done by dropping one

colour and picking up the other, using the right hand. If you are lucky enough to have mastered both the "English" and "Continental" methods of knitting, the yarns being used may be held simultaneously, one in the left hand, the other in the right hand. The instructions below, however, are limited to the more standard one-handed method and give the three alternative methods of dealing with the yarn not in use.

Stranding

Stranding is the term used to describe the technique by which the yarn not in use is simply left hanging at the back of the work until it is next needed. The yarn in use is then dropped and the carried yarn taken up, ready for use. This means that the strand, or "float", thus produced on the wrong side of the work has a direct pull on the stitches either side of it.

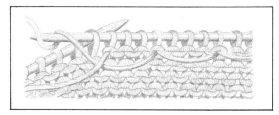

It is essential to leave a float long enough to span this gap without pulling the stitches out of shape and to allow the stitches in front of it to stretch and not to pucker on the right side of the work. It is preferable to go to the other extreme and leave a small loop at the back of the work rather than pull the float too tightly.

If the gap to be bridged by the float is wide, the strands produced may easily be caught and pulled when the garment is put on or taken off. This problem may be

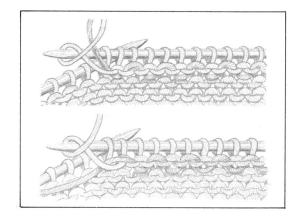

12

remedied by catching the floats down with a few stitches on the wrong side of the work at the finishing stage.

Weaving

With this method the yarn being carried is looped over or under the working yarn on every stitch, creating an up and down woven effect on the wrong side of the work. Since the knitter does not have to gauge the length of the floats, many people find that this is the easiest method of ensuring an even, accurate tension. Weaving does increase the chances of the carried colour showing through on to the right side of the work, however, and it tends to produce a far denser fabric, which is not always desirable when a thick fibre is being used.

Stranding and weaving

Combining the two methods of stranding and weaving is invariably the most practical solution to the problem of working perfect fairisle. Most designs will have colour areas that will vary in the number of stitches. If the gap between areas of the same colour is only a few stitches then stranding will suffice, but if the float produced will be too long, weave the carried yarn in every few stitches. Should you be unsure about the length of float to leave, slip your fingers under one. If you succeed with ease, the float is too long.

The most difficult aspect of fairisle knitting is getting the tension correct. This does not depend on the stitch size so much as on the way you treat the carried yarn. This is why, when working an all-over fairisle, you should always knit a tension sample in fairisle, not in main colour stocking stitch, as the weaving or stranding will greatly affect the finished measurement of the stitches. The most important rule to remember is that *the yarn being carried must be woven or stranded loosely enough to have the same degree of "give" as the knitting itself.* Unless this is achieved, the resulting fabric will have no elasticity whatsoever and, in extreme examples very tight floats will buckle the stitches so that they lie badly on the right side of the work.

If you are using the fairisle technique to work a colour motif on a single-colour background, keep the motif tension as close to the background tension as possible. If there is a great difference, the motif stitches will distort the image.

INTARSIA

Intarsia is the term used for the technique of colour knitting whereby each area of colour is worked using a separate ball of yarn, rather than carrying yarns from one area to another as in the fairisle technique. Any design that involves large blocks of isolated colour that are not going to be repeated along a row or required again a few rows later, should be worked in this way.

There are no limitations to the number of colours that may be used on any one row other than those imposed by lack of patience and/or dexterity. Avoid getting into a tangle with too many separate balls of yarn hanging from the back of the work and remember that every time a new ball of yarn

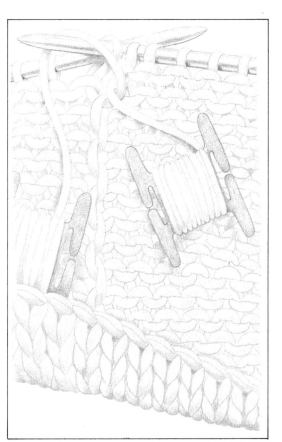

Left: if you are using the intarsia method, twist the yarns firmly together when you change colours.

Far left: stranding and weaving worked too tightly.

is introduced and broken off after use, two extra ends are produced that will have to be secured at the end of the day. When ends are left, always make sure that they are long enough to thread up so that they may be properly fastened with a pointed tapestry needle. Do this very carefully through the backs of the worked stitches to avoid distorting the design on the right side of the work. The ends that are left should never be knotted because they will make the wrong side of the work look extremely unsightly and they will invariably work themselves loose and create problems at a later stage.

If only a few large, regular areas of colour are being worked, avoid tangling the wool by laying the different balls of yarn on a table in front of you or keep them separate in individual jam jars or shoe-boxes. However, this requires careful turning at the end of every row so that the various strands do not become twisted.

The easiest method is to use small bobbins that hold each yarn separately and that hang at the back of the work. Such bobbins are available at most large yarn stores or they may be made at home out of stiff card. They come in a variety of shapes, but all have a narrow slit in them that keeps the wound yarn in place but allows the knitter to unwind a controlled amount as and when required. When winding yarn on to a bobbin, try to wind sufficient to complete an entire area of colour, but don't overwind, as heavy bobbins may pull stitches out of shape.

When you change colour from one stitch to another, it is essential that you twist the yarns around one another before dropping the old colour and working the first stitch in the new colour. This prevents a hole from forming. If it is not done, there is no strand to connect the last stitch worked in colour "A" to the first stitch worked in colour "B". This twisting should also be done quite firmly to prevent a gap from appearing after the work has settled.

CASTING ON

The methods of casting on are too numerous to mention and choosing one is normally very much a matter of personal preference. When you are using cotton, however, the yarn itself must dictate the method used, because some methods will produce a cast-on edge that is far too uneven or "frilly". The technique described here will be familiar, in part, to most knitters since it involves the simple thumb method of casting on. By itself, this can produce an untidy edge, but with the addition of knitting into each stitch formed on the thumb, a neat, firm cast-on edge will be produced.

When you start to cast on, do not make a slip loop and put it on the needle, as is sometimes recommended, because this

A neat cast-on edge will result if you use the thumb method and, in addition, knit into each stitch.

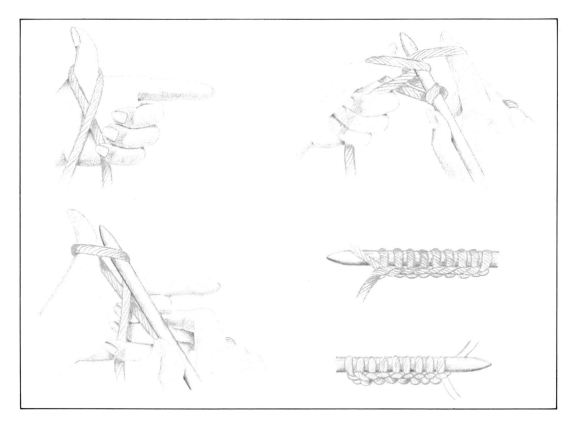

forms a noticeable knot when the work is completed. Instead, start straight in on the casting-on method, forming the first loop on the needle by slipping it off the thumb after step 2. When you come to form the next stitch, support the first loop at the back of the needle with the index finger of your left hand as shown. This will stop the yarn twisting around the needle while you make the first proper stitch by knitting into the loop you have made around your thumb. These three steps are repeated until the required number of stitches has been formed.

The work may now start straight away with row 1, which will be a wrong side (WS) row since it is the "knot" side of the cast-on stitches.

MAKING A BOBBLE

There are numerous variations on the theme of bobble making, but in this book we have used just two, which, for ease of identification, we have called large and small, abbreviated as MBL and MBS. If worked on a right side (RS) row, the bobble will hang on the right side, if worked on a wrong side (WS) row, push it through on to the right side.

Large bobble

1. When the MBL position on the row has been reached, make 5 stitches out of the next one by knitting into its front, then its back, front, back and front again before slipping it off the LH needle.
2. Turn the work and knit these 5 stitches only.
3. Turn the work, purl 5 and repeat the last 2 rows.
4. Using the point of the left-hand needle, lift the bobble stitches, in order, over the first one on the right-hand needle, i.e., 2nd, 3rd 4th and 5th, so that one stitch remains.

After completing the bobble, the work may continue as normal, the single stitch having been restored to its original position on the row.

Small bobble

This is made in exactly the same way as the large bobble, except that 4 stitches are made out of the 1, instead of 5. Rows 2 and 3 are worked but are not repeated.

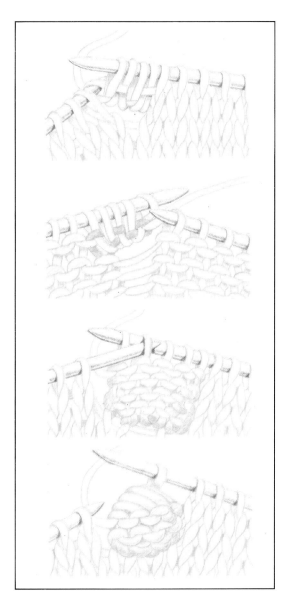

The four steps in making a bobble are illustrated here. If you work a bobble on the wrong side (WS), push it through to the right side (RS) of your work.

TURNING

Turning is also called "working short rows" since this is precisely what is being done. By turning the work in mid-row and leaving part of it unworked, shaping is created since one side of the work will have more rows than the other. If the work is then cast off, a sloping edge will result, making this method ideal for shoulder shaping instead of the more usual method of casting off groups of stitches to produce noticeable "steps". Turning is not advisable if you are working complicated stitches or colour patterns as working short rows will invariably throw them out.

Unfortunately, even when care is taken, there is a tendency for holes to form at the points where the work is turned. There is a method whereby these holes may be

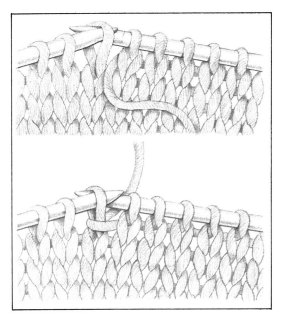

The first steps in turning to ensure that holes do not form.

completely eliminated, and although this looks complicated at first, the finished effect is well worth the effort involved in mastering the technique.

This method may be used on right or wrong side rows. Here it has been illustrated on the right side of stocking stitch.

1. Knit to the point where turning is indicated, but before doing so bring the yarn to the front of the work and

slip the next stitch from the left-hand to the right-hand needle.
2. Put the yarn to the back of the work and return the slipped stitch to the left-hand needle.
3. Now turn the work and purl to end.

Repeat the last three steps at every turning point.

All the stitches must now be worked across, so if the turned shaping is being worked immediately before casting off or knitting a seam, add an extra row. Work to the first stitch that has had a loop made around it by putting the yarn forward and then back.

1. Slip this stitch from the left-hand to the right-hand needle, lifting the strand of the loop on to the right-hand needle along with the stitch.
2. Slip both the strand and the stitch back to the left-hand needle, straightening them as you do so.
3. Knit the stitch and strand together.

Work to the next "looped" stitch and repeat the process.

When this row is completed, the work may be continued as normal.

CABLES

A basic cable is simply a twist in the knitted fabric caused by working a small number of stitches out of sequence every few rows. This is done by slipping the stitches on to a needle and leaving them at the front or the back of the work while the next stitches on the left-hand needle are worked. The held stitches are then worked normally. The cable, worked in stocking stitch, will always be flanked by a few reversed stocking stitches to give it definition. Since it does involve a twist, however, cabled fabric will always have a tighter tension than one worked in plain stocking stitch, so take extra care when working a tension sample.

Cable needles are very short and double-ended. Some have a little kink in them to help keep the stitches in place while others are being worked. Use one that is a similar size to the needles being used for the main work and take care not to stretch or twist the stitches when moving them from needle to needle.

On the right side of the work, if the

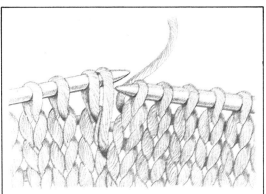

The final steps of turning.

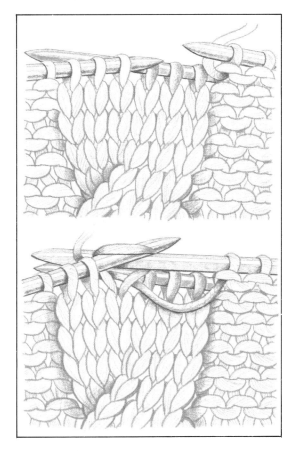

stitches are held to the front, the cable will cross from the right to the left. If the stitches are held at the back of the work, the cable will twist from the left to the right.

Front cross cable

1. (RS): work to the six stitches that are to be cabled. Slip the next three stitches on the left-hand needle on to the cable needle and leave them hanging at the front of the work.
2. Knit the next three stitches on the left-hand needle as normal.
3. Knit the three held stitches off the cable needle.

Repeat this twist wherever indicated in the instructions.

The same basic technique may be used to move a single stitch across a background of stocking stitch at a diagonal, rather than form a cable that moves up the work vertically.

Where the abbreviation cb2 is used, the first stitch is slipped on to the cable needle and left at the back of the work while the next stitch is knitted. The held stitch is then knitted off the cable needle. Cf2 is the same but with the cable needle left at the front of the work. On purl rows the abbreviations pcb2 and pcf2 are used to denote the same movement but in which the stitches are purled rather than knitted. In this way a continuous criss-cross line is formed.

SEAMS

After achieving the correct tension, the final sewing up of your knitting is the most important technique to master. It can make or break a garment, however carefully it may have been knitted. This is why the making up instructions after every set of knitting instructions should be followed exactly, especially to the type of seam to be used and the order in which the seams are to be worked.

Before starting any piece of work, always leave an end of yarn long enough to complete a substantial section, if not the whole length, of the eventual seam. After working a couple of rows, wind this up and pin it to the work to keep it out of the way. If required, also leave a sizeable end when the work has been completed. This saves having to join in new ends that may well work loose, especially at stress points such as welts.

The secret of perfect-looking seams is uniformity and regularity of stitch. When joining two pieces that have been worked in the same stitch, they should be joined row for row, and all work should be pinned first to ensure an even distribution of fabrics. When joining work that has a design on both pieces, take great care to match the colours, changing the colour you are using to sew the seam where necessary.

Backstitch

Pin the two pieces of work together, right sides facing, making sure that the edges are absolutely flush. Always leave as narrow a seam allowance as possible to reduce unnecessary bulk. It is essential that the line of backstitches is kept straight, using the lines of the knitted stitches as a guide. All the stitches should be identical in length, one starting immediately after the previous one has finished. On the side of the work facing you, the stitches should form a continuous, straight line. If the seam is starting at the very edge of the work, close the edges with an overstitch as shown. Now work the backstitch as follows:

The front cross cable.

If you use backstitch to join a seam and you are starting at the very edge of the work, close the edges with an overstitch before beginning the row of backstitch.

When you use backstitch to join a seam, the finished seam should be perfectly straight; the drawings illustrate the appearance of the front (top) and the back (below) of the completed seam.

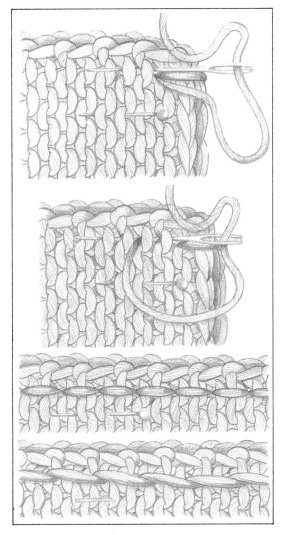

1. Make a running stitch (maximum length 1cm), through both thicknesses of work.
2. Put the needle back into the work in exactly the same spot as before and make another running stitch twice as long.
3. Put the needle back into the work adjacent to the point where the previous stitch ended. Make another stitch the same length.

Keep repeating stage 3 until the last stitch, which needs to be half as long to fill in the final gap left at the end of the seam.

By keeping the stitch line straight and by pulling the yarn fairly firmly after each stitch, no gaps should appear when the work is opened out and the seam pulled apart.

This seam is suitable for lightweight

The drawings show how you should hold the knitting to work a flat seam and how the work will look on the right side.

yarns or when an untidy selvedge has been worked.

Flat seam

This seam is a slight contradiction in terms since its working involves an oversewing action, but when the work is opened out it will do so completely and lie quite flat, unlike a backstitched seam.

Use a blunt-ended tapestry needle to avoid splitting the knitted stitches. Pin both pieces right sides together and hold the work as shown. The needle is placed through the very edge stitch on the back piece and then through the very edge stitch on the front piece. The yarn is pulled through and the action repeated, with the needle being placed through exactly the same part of each stitch every time. Always work through the edge stitch only. By taking in more than this, a lumpy, untidy seam that will never lie flat will be produced.

When two pieces of stocking stitch are to be joined with a flat seam, do not work any special selvedge such as knitting every edge stitch. Just work the edge stitches normally but as tightly as possible, using

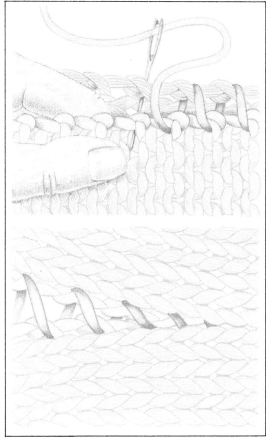

only the tip of your needle. When you come to work the seam, place the needle behind the knots of the edge stitches and not the looser strands that run between the knots, since these will not provide a firm enough base for the seam, which will appear gappy when opened out.

Flat seams are essential for heavy-weight yarns where a backstitch would create far too much bulk. They should also be used for attaching buttonbands, collars and so forth, where flatness and neatness are essential.

Borders, welts, cuffs and any other part

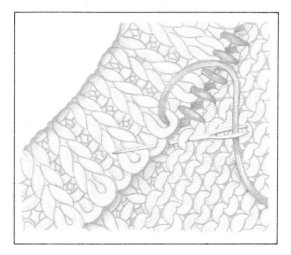

of a garment where the edge of the seam will be visible should be joined with a flat seam, even if the remainder of the garment is to have a backstitched seam. Start with a

flat seam until the rib/border is complete and then change over to a backstitch, taking in a tiny seam allowance at first and then smoothly widening it without making a sudden inroad into the work.

Slip stitch
Where one piece of work is to be placed on top of another, for example when turning in a double neckband, folding over a hem or attaching the edges of pocket borders, a slip stitch should be used.

When turning in a neckband that has been cast off, the needle should be placed through the cast-off edge and then through the same stitch but on the row where it was initially knitted up. It is essential to follow the line of the stitch in this way to avoid twisting the neckband. By repeating the action, the visible sewn stitch runs at a diagonal.

The same rule applies when sewing down a neckband that has not been cast off but that has had its stitches held on a thread. The only difference is that the needle is placed through the actual held stitch, thus securing it. When each stitch has been slip stitched down, the thread may be removed. This method allows for a neckband with more "give" than one that has been cast off.

On pocket borders, use the line of stitches on the main work as a guide to produce a perfectly straight vertical line of stitches. Place the needle through one strand of the main work stitch and then behind the knot of the border edge stitch, as for a flat seam.

KNITTED SHOULDER SEAMS
This method of joining is perfect for shoulders on which no shaping has been worked or on which the shaping has been worked by turning rows, as described on pages 15-16. It creates an extremely neat, flat seam.

Since the two pieces to be joined must be worked stitch for stitch, they must both have exactly the same number of stitches. Even though the pattern specifies that you should have a certain number of stitches at this point, it is wise to double check the number you actually have on your needles, since it is very easy to lose or gain the odd stitch accidentally along the way.

The technique itself involves the use of three needles. The stitches from the front

Use slip stitch to hold a double neckband in position and (left below) to attach a pocket to a garment.

The diagrams illustrate the three steps involved in knitting shoulder seams together. You must always have exactly the same number of stitches on the two pieces that are to be joined in this way.

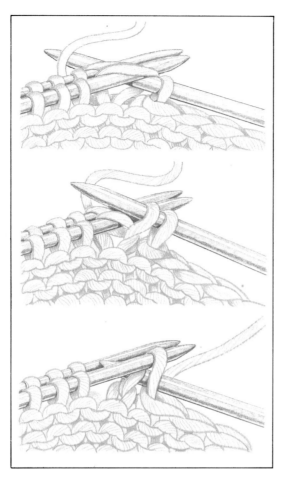

right-hand needle. The second stitch is then lifted over the first, as in normal casting off.

Step 3 should be repeated across all the stitches to be knitted together until one loop remains on the right-hand needle. Pull the yarn through this to secure.

When knitting together a shoulder seam on a garment where no neck shaping has been worked and the neck stitches have not been cast off, the back stitches may be dealt with altogether. By starting to work the first shoulder together from the armhole to the neck edge, the back neck stitches may then be cast off (if the pattern requires that they are cast off), without breaking the yarn, which may then be used to knit together the second shoulder seam.

Normally worked on the inside of the work to create an extremely neat, flat and durable seam on the right side of the work, a knitted seam may be worked with the wrong sides of the knitting facing one another. This creates a decorative ridge on the right side of the work.

and back are held on their respective needles, both of which are held in the left hand, while the right hand holds a third, larger, needle, which helps to prevent the cast-off stitches from becoming too tight. Holding more than one needle in the hand and trying to work through two stitches at a time without dropping them can seem awkward at first, but, with a little practice, it will feel like normal knitting.

The needles are held with the right sides of the work facing one another and with the stitches lined up at corresponding intervals on the front and back needles.

1. The point of the right-hand needle is put through the first stitch on the front needle and the first stitch on the back needle, with exactly the same action as a normal kit stitch but going through both simultaneously.
2. A loop is pulled through to form a single stitch on the right-hand needle, the old stitches being slipped off the left-hand needle.
3. These two steps are repeated so that there are two stitches on the

When Swiss darning, use a yarn of the same thickness as the knitting and follow the path of the knitted stitches.

SWISS DARNING

This is the most straightforward method of embroidery that may be worked on knitted fabrics since it exactly replicates knitted stitches. For this reason Swiss darning is sometimes called "duplicate stitch". By following the path of the knitted stitch with a contrasting colour, it is possible to create a variety of designs that have the appearance of being worked as a complicated fairisle, although they have, more simply, been added afterwards. For knitters who are not too confident with colour techniques, this is a very useful adjunct to their knitting skills.

When Swiss darning, always use a yarn of the same thickness as the knitting so that it will cover the stitch beneath it but not create an embossed effect. Use a blunt-ended tapestry needle to avoid splitting the knitted stitches as you embroider. The tension of the embroidered stitches must be kept exactly the same as the work that is providing the base so that they sit properly and do not pucker the work. The tension is regulated by how tightly the embroidery yarn is pulled through the work at each stage. Take great care when joining in and securing yarn ends on the wrong side of the work so that the stitches in the area do not become distorted.

The illustrations show the path of the embroidery yarn which will create the diagonal lines required on the Diamond-front Cardigan (page 50). The same method may be used to create horizontal and vertical lines, and you should always use the path of the knitted stitch as your guideline.

TWIN-SET
WITH LACE TRIM

A sleeveless sweater and classic cardigan combination worked in a subtle combination of white and ecru 4-ply mercerized cotton. The sweater has a knitted lace collar, which is worked in two pieces, allowing an opening at the back which is fastened with a pearl button. The lace is echoed on the pocket trim of the cardigan. Sizes quoted are women's small/medium/large.

Materials
Melinda Coss 4-ply mercerized cotton –
Sweater ecru: 250/250/300gm; white: 50gm;
1 pearl button. **Cardigan** ecru:
400/450/500gm; white: 50gm; 8 pearl
buttons.

Needles
One pair of 2¾mm and one pair of 3¼mm
needles.

Tension
Using 3¼mm needles and measured over
st st, 30 sts = 10cm.

SWEATER
Front
Using 2¾mm needles and ecru, cast on
128/134/140 sts.
Row 1 (WS): *k1, p1, rep from * to end. Row
2: *k1 tbl, p1, rep from * to end. Keep rep
these 2 rows to form twisted rib for 10cm,
ending with a WS row. Change to 3¼mm
needles.
Row 1: k1, inc into next st, k10, inc into next
st, k to last 12 sts, inc into next st, work to
last 2 sts, inc into next st, k1. Cont in st st
until the work measures 31/32/33cm.

Shape armholes: cast off 12 sts at beg of next 2 rows. Now dec 1 st each end of every row for 10/10/13 rows and then every alt row until 78/84/84 sts remain. Work straight until the front measures 46/48/50cm, ending with a WS row.

Shape neck: next row: k 30/33/31 sts, cast off 18/18/22 sts, k to end. Cont with this set of sts, leaving others on a holder. Now dec 1 st at neck edge on every row for 10 rows (20/23/21 sts). Work 3 rows straight.

Shape shoulder: *see* Techniques, pages 15–16. Rows 1 and 2: k to last 7 sts, turn and p to end. Rows 3 and 4: k to last 14 sts, turn and p to end. Knit across all sts, leave on a holder. Return to other set of sts, join yarn in at neck edge and work to match first side, reversing out the shapings.

Back

Work as for front until work measures 42/44/46cm, ending with a WS row.

Divide for opening: next row: k36/39/39 sts, put these sts on a holder and k to end of the remainder. Next row: p to last 6 sts, k6. Cont thus in st st with a 6 st garter st (knit every row) border until work measures 45/47/50cm, ending with an RS row.

Work buttonhole: work to last 4 sts, cast off 2, k to end. Row 2: k, casting on 2 sts above those cast off on previous row. Row 3: p to last 6 sts, k6.

Shape neck: cast off 16 sts at beg of next row. Now dec 1 st at neck edge on every row until 20/23/21 sts remain. Work 1 row straight, then **shape shoulder** as for front. Work across all sts and leave on a holder.

Cast on 6 sts and join these in at the point where sts were divided and p to end of held sts. Cont as for other side but omitting buttonhole. Work neck and shoulders to match first side, reversing out the shapings. Knit both shoulder seams tog.

Left armband

Open work out and using ecru, 2¾mm needles and with RS facing, knit up 164/172/180 sts from front to back. Purl the first row and then cont in twisted rib as for welts. When rib measures 2cm, work one more row, then cast off in rib using a larger needle.

Right armband

Work as for left armband but knitting up sts from back to front.

Collar

(Worked in 2 pieces.)
Using 3¼mm needles and white, cast on 8 sts.

N.B. The yo at beg of every RS row is intentional and forms an attractive little loop edging.

Row 1 (RS): yo, k2 tog, k1, yo, k to end.
Row 2 (and all WS rows): k to last st, p1.
Rows 3 and 5: yo, k2 tog, k1, yo, k2 tog, yo, k to end.
Row 7: yo, k2 tog, k1, (yo, k2 tog) twice, yo, turn work.

Row 8: k to last st, p1.
Row 9: yo, k2 tog, k1, (yo, k2 tog) twice, yo, k to end.
Rows 11 and 13: yo, k2 tog, k1, (yo, k2 tog) 3 times, yo, k to end.
Row 15: yo, k2 tog, k1, (yo, k2 tog) 4 times, yo, turn work.
Row 16: k to last st, p1.
Row 17: yo, (k2 tog) twice, (yo, k2 tog) 4 times, k to end.
Rows 19 and 21: yo, (k2 tog) twice, (yo, k2 tog) 3 times, k to end.
Row 23: yo, (k2 tog) twice, (yo, k2 tog) twice, k1, turn work.
Row 24: k to last st, p1.
Row 25: yo, (k2 tog) twice, (yo, k2 tog) twice, k to end.
Rows 27 and 29: yo, (k2 tog) twice, yo, k2 tog, k to end.
Row 31: yo, (k2 tog) twice, k1, turn work.
Row 32: k to last st, p1.
These 32 rows form the pattern.
Rep pattern 3 more times. Cast off. Work another collar piece to match.

Making up
Slip st down the buttonband at its base. Attach button. Join side seams with a flat seam over the ribs, a narrow backstitch over the st st.
Attach collar pieces (taking care to centre them), with a flat seam.

CARDIGAN

Back
Using 2¾mm needles and ecru, cast on 140/148/158 sts and work in twisted rib for 10cm, ending with a WS row. Next row: k, inc into every 14th/14th/15th st (150/158/168 sts). Change to 3¼mm needles and cont in st st until work measures 33/34/35cm.
Shape armholes: cast off 4 sts at beg of next 2 rows. Now dec 1 st each end of every row for 8 rows. Now dec 1 st each end of every alt row until 114/122/132 sts remain. Work straight until back measures 53/55/57cm, ending with a WS row.
Shape neck: k36/39/43 sts, cast off 42/44/46 sts, k to end. Cont with this set of sts, leaving the others on a holder.
Next row: p to last 3 sts, p2 tog, p1.
Rows 2 and 3: k to last 11/12/13 sts, turn work, p to last 3 sts, p2 tog, p1.
Rows 4 and 5: k to last 22/24/26 sts, turn work, p to end. Leave sts on a holder.

Return to other sts, joining yarn in at armhole edge.
Next row: k to last 3 sts, k2 tog, k1.
Rows 2 and 3: p to last 11/12/13 sts, turn work, k to last 3 sts, k2 tog, k1.
Rows 4 and 5: p to last 22/24/26 sts, turn work and p to end. Leave sts on a spare needle/holder.

Left front
Before starting front, work a pocket lining by casting on 21 sts, using 3¼mm needles and ecru. Work in st st for 6cm, ending with an RS row. Leave sts on a spare needle.
Using 2¾mm needles and ecru, cast on 74/78/82 sts and work in twisted rib for 10cm, ending with an RS row.
Next row: rib 8 and place these sts on a safety-pin, p to end, inc into every 9th/10th/10th st (73/77/81 sts). Change to 3¼mm needles and cont in st st until work measures 33/34/35cm, ending with a WS row.
Shape armholes: cast off 4 sts at beg of next row. Now dec 1 st at armhole edge on the next 8 rows (61/65/69 sts). Cont, dec 1 st at armhole edge on every alt row until 55/59/63 sts remain. Now cont working straight.
Next RS row: k17/19/21 sts, cast off 21 sts and k to end.
Row 2: p to cast off sts and p the pocket lining sts instead, p to end.
Cont working straight until work measures 48/50/52cm, ending with an RS row.
Shape neck: cast off 7 sts at beg of next row. Now dec 1 st at neck edge on every row until 34/37/41 sts remain. Work straight until work measures 53/55/57cm, ending with an RS row. Next row: p to last 11/12/13 sts, turn work, k to end.
Next row: p to last 22/24/26 sts, turn and k to end. Leave these sts on a holder.

Right front
As for left front until 3 rows of rib have been worked.
Work buttonhole: next row (RS): rib 3, cast off 2, rib to end. Row 2: rib to end, casting on 2 sts above those cast off on previous row. Cont in rib, working another buttonhole when work measures 6cm. Now cont as for left front, reversing out shaping and omitting breast pocket.

Sleeve
Using 2¾mm needles and ecru, cast on 50/54/54 sts and work in twisted rib for

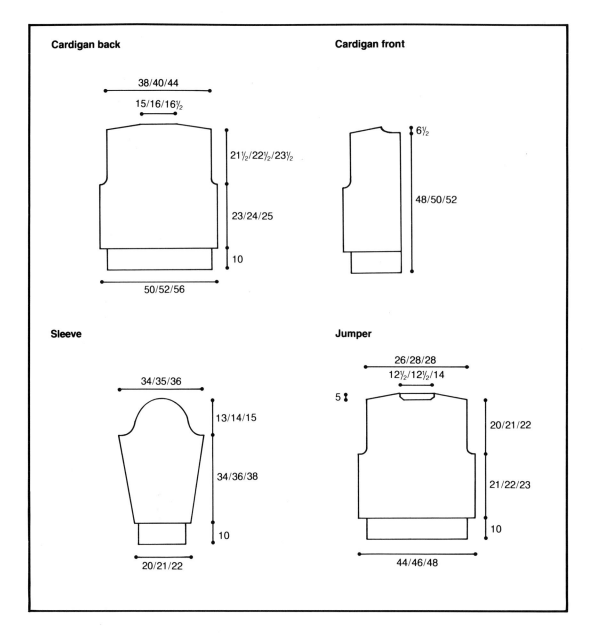

Cardigan back

38/40/44

15/16/16½

21½/22½/23½

23/24/25

10

50/52/56

Cardigan front

6½

48/50/52

Sleeve

34/35/36

13/14/15

34/36/38

10

20/21/22

Jumper

26/28/28

12½/12½/14

5

20/21/22

21/22/23

10

44/46/48

10cm, ending with a WS row. Next row: knit, inc into every 5th/5th/6th st (60/64/67 sts). Change to 3¼mm needles and cont in st st, inc 0/0/1 st at beg of next row. Now inc 1 st at each end of the next/next/6th row and every following 5th/5th/6th row until you have 102/106/110 sts. Now work straight until the sleeve measures 44/46/48cm from the beg.

Shape sleeve head: cast off 4 sts at beg of next 2 rows. Now dec 1 st each end of every row until 48/40/32 sts remain. Now cast off 4 sts at the beg of every row until 8 sts remain. Cast these off.

Left band

Slip the 8 sts off the safety-pin on to a 2¾mm needle and work in twisted rib until the

band reaches the point where the neck shaping starts when very slightly stretched. Put back on to a pin and mark a button position 5cm down from the last row worked. Now mark five other button positions between this and the second buttonhole on the buttonhole band.

Right band

Work as for left band, working buttonholes to correspond to the button positions which have been marked.

Neckband

Knit both shoulder seams tog (*see* Techniques, pages 19–20). Using 2¾mm needles, ecru and with RS facing, slip the right band sts on to the needle and then knit

up 25 sts from the right front, 50/52/54 sts across the back neck, 25 sts down the left front and then slip the right band sts on to the needle (116/120/124 sts). Work in twisted rib for 5 rows. Next row: rib 3, cast off 2, rib to end. Cont in rib, casting on 2 sts above those cast off on previous row. When neckband measures 2cm, rib 1 more row and then cast off in rib.

Pocket trim

Using 3¼mm needles and white, cast on 8 sts and work in lace pattern as for sweater collar for 32 rows. Cast off.

Making up

Slip st pocket lining in position on WS of work, taking care that the sts do not show through on to RS of work. Put trim into pocket so that the garter sts are not showing. Slip st the cast-off edge of the pocket front to the trim, which then folds over it to look like a hanky point.

Join the side and sleeve seams with a flat seam over the ribs and a backstitch over the st st (*see* Techniques, pages 17–18). Pin sleeves into armholes, taking care to distribute them evenly. Join with a backstitch. Attach buttons where indicated.

BUTTON-THROUGH SWEATER

A drop-shoulder, loose-fitting sweater quoted in women's sizes 1/2. The slub-spun cotton which is used gives an interesting texture to a garment worked in stocking stitch. Detailing includes ornamental knitted seams and seven pearl buttons.

Materials
Melinda Coss slub cotton – 500/550gm; 7 pearl buttons.

Needles
One pair of 3mm and one pair of 3¼mm needles.

Tension
Using 3¼mm needles and measured over st st, 26 sts and 34 rows = 10cm square.

Back
Using 3mm needles, cast on 140/152 sts. Row 1:*k1, p1, rep from * to end. Keep rep this row to form single rib for 6cm. Change to 3¼mm needles and start working in st st, inc 5 sts equally across the first row (145/157 sts). Cont in st st until the work measures 68/73cm. Leave sts on a spare needle.

Front
Work as for back until it measures 39/44cm, ending with a WS row.
Divide for opening: next row, k69/75 sts, cast off 7 sts, work to end. Cont with this set of sts, leaving the others on a holder, until work measures 60/65cm, ending with a WS row.
Shape neck: cast off 7 sts at beg of row, then dec 1 st at neck edge on every row for 7 rows and then on every alt row until 48/54 sts

remain. Now work straight until front
matches back. Leave sts on a holder. Return
to the other sts and work the other side of
the neck to match. Leave sts on a holder.

Sleeves

Using 3mm needles, cast on 50/54 sts and
work in single rib for 6cm, ending with a WS
row. Knit the next row, inc into every
12th/5th st (54/64 sts). Change to 3¼mm
needles and cont in st st, inc 1 st each end of
the 3rd/next row and every following
3rd/4th row until you have 120/124 sts. Work
straight until sleeve measures 39/42cm.
Leave sts on a spare needle.

Buttonband

With RS of left front facing and using a 3mm
needle, knit up 44 sts from the point where
the work was divided to the neck edge. Purl
the first row (this forming a ridge on RS of
the work). Now work in single rib for 2½cm,
ending with an RS row. Cast off, knitwise,
using a larger needle to keep the tension of
the edge the same as the ribbing.

Buttonhole band

Knit up as for buttonband but working from
the neck edge to the point where the work
was divided. Purl the first row and then
work 3 rows in rib. Now work the
buttonhole rows.
Row 1: *rib 5, cast off 2, rep from * 5 more
times, rib 2.
Row 2: rib, casting on 2 sts above those cast
off on the previous row.
Complete as for other band.

Neckband

Knit both shoulder seams tog, with WS
facing so that a decorative ridge is formed on
RS of the work (*see* Techniques,
pages 19–20). Using 3mm needles and with
RS facing, knit up 14 sts from buttonband
and cast-off edge on left front, 21 sts from the
side of the neck. Slip the 49 back neck sts on
to the needle and then knit up 21 sts down
other side and 14 sts from the right front (119
sts). Purl the first row and then cont in rib
for 2cm. On the next row work a 2 st
buttonhole in line with those previously
worked. Cont until rib is 2½cm deep, ending
with an RS row. Cast off knitwise, using a
larger needle as before.

Making up

Attach the bottom edge of the buttonhole

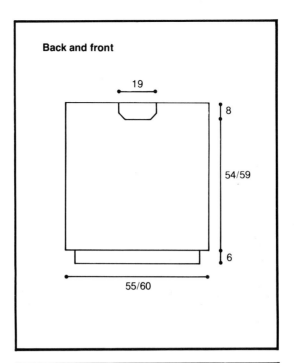

Back and front

19

8

54/59

6

55/60

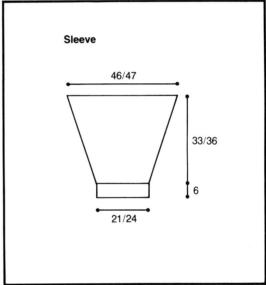

Sleeve

46/47

33/36

6

21/24

band to the cast-off front edge using a flat
seam. Slip st the buttonband edge
underneath this. Attach the buttons to
correspond to the buttonholes.
Open out the front and back and mark the
armholes, 24/25cm either side of the
shoulder seam. Using 3mm needles and
with RS facing, knit up 120/124 sts from
marker to marker, taking great care to
distribute them evenly. Now hold WS of
body and sleeve tog and knit these sts tog so
that they form a ridge on the outside of the
work, as on shoulder.
Join sleeve and side seams using a flat seam
over ribs and a narrow backstitch over the
st st (*see* Techniques, pages 17-18).

BOBBLE-FRONTED CARDIGAN

A fitted Forties-shaped cardigan in 4-ply mercerized cotton, worked in three sizes to fit bust sizes 82/87/92cm. The 12 buttons mean that, when fastened, the cardigan can easily be worn as a sweater. The fancy panels involve a simple open stitch surrounded by bobbles (*see* Techniques, page 15).

Materials
Melinda Coss 4-ply mercerized cotton – 350/400/400gm; all sizes: 12 buttons.

Needles
One pair each of 2¾mm, 3mm and 3¼mm needles.

Tension
Using 3¼mm needles and measured over st st, 30 sts = 10cm.

Back
Using 3mm needles, cast on 108/116/124 sts. Row 1: *k2, p2, rep from * to end. Keep rep this row to form double rib. Work for 5cm, then change to 2¾mm needles and work until rib measures 9cm from beg. Change to 3¼mm needles and cont in st st until work measures 10cm, ending with a WS row.
Shape dart: row 1: k30/32/34 sts, m1, k to last 30/32/34 sts, m1, k to end. Now work 7 rows without shaping. Rep these last 8 rows, 8 more times (126/134/142 sts). Work straight until work measures 32/33/34cm from beg.
Shape armholes: cast off 8/9/10 sts at beg of next 2 rows. Now dec 1 st each end of next and every alt row until 98/102/106 sts remain. Work straight until armhole measures 18/19/20cm from beg of shaping, measured on the straight.

Shape shoulders: cast off 10 sts at beg of next 4 rows. Now cast off 9/10/11 sts at beg of next 2 rows. Leave remaining sts on a holder.

Right front

Using 3mm needles, cast on 56/60/64 sts and work in double rib, as for back (changing needle size at the same point), until work measures 9cm, ending with a WS row. Change to 3¼mm needles and work 6 rows in st st. Now start working the fancy panels.

Row 1: k25/27/29 sts, MBS, k to last 2 sts, M1, k to end.
Row 2: (and every WS row): purl.
Row 3: k23/25/27 sts, MBS, k1, yfwd, k2 tog, MBS, k to end.
Row 5: k21/23/25 sts, MBS, (k2 tog, yfwd) 3 times, k1, MBS, k to end.
Row 7: k19/21/23 sts, MBS, k1, (yfwd, k2 tog) 5 times, MBS, k to last 2 sts, M1, k to end.
Row 9: k17/19/21 sts, MBS, (k2 tog, yfwd) 7 times, k1, MBS, k to end.
Row 11: k15/17/19 sts, MBS, k1, (yfwd, k2 tog) 9 times, MBS, k to end.
Row 13: k17/19/21 sts, MBS, (k2 tog, yfwd) 7 times, k1, MBS, k to last 2 sts, M1, k to end.
Row 15: k19/21/23 sts, MBS, k1, (yfwd, k2 tog) 5 times, MBS, k to end.
Row 17: k21/23/25 sts, MBS, (k2 tog, yfwd) 3 times, k1, MBS, k to end.
Row 19: k23/25/27 sts, MBS, k1, yfwd, k2 tog, MBS, k to last 2 sts, M1, k to end.
Row 20: purl.

These 20 rows form the pattern, with an increase being worked at the seam edge on

every 6th row. Cont in pattern, inc as before until you have 64/68/72 sts. Now cont straight, in pattern, until work measures 32/33/34cm from beg, ending with an RS row.

Shape armhole: cast off 8/9/10 sts at beg of next row. Now dec 1 st at this edge on every alt row until 47/49/51 sts remain. Now cont straight (still in pattern until 7 diamonds

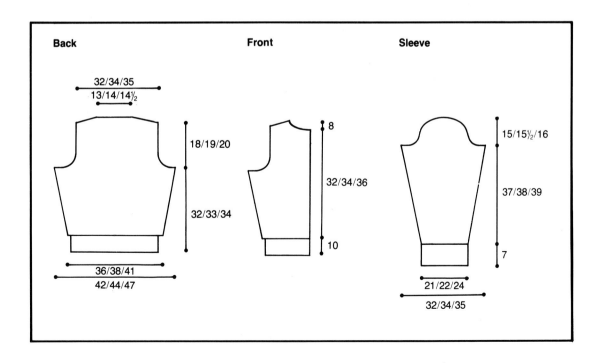

have been completed and from then on in st st), until work measures 10/11/12cm from the start of the armhole shaping, ending with a WS row.

Shape neck: cast off 8/9/10 sts at beg of next row and then dec 1 st at this edge on every alt row until 29/30/31 sts remain. Work this edge straight, until work measures 18/19/20cm from the start of the armhole shaping, ending with an RS row.

Shape shoulder: cast off 10 sts at beg of next 2 WS rows. Work 1 row straight, then cast off remaining sts.

Left front

Work as for right front, reversing out the shapings.

Sleeves

Using 3mm needles, cast on 56/60/64 sts and work in double rib for 7cm, ending with a WS row. Next row: knit, inc into every 7th/7th/8th st (64/68/72 sts). Change to 3¼mm needles and cont in st st, inc 1 st each end of every 8th row until you have 98/102/106 sts. Now work straight until sleeve measures 44/45/46cm.

Shape sleeve head: cast off 7 sts at beg of next 2 rows. Now dec 1 st each end of every row until 68/72/76 sts remain. Now dec 1 st each end of every alt row until 28/30/32 sts remain. Cast off 4/5/6 sts at beg of next 4 rows. Cast off remaining sts.

Buttonband

Using 3mm needles, cast on 10 sts and work in double rib until the band is long enough to reach from the bottom edge to the start of the neck shaping when very slightly stretched. Leave sts on a safety-pin. Mark 11 button positions (allowing for the 12th, which will go on the neckband), with pins.

Buttonhole band

Work as for the buttonband, but working a buttonhole to correspond with the pins, marking the button positions thus: Rib 4 sts, cast off 2 sts, rib to end. Next row: rib, casting on 2 sts above those cast off on previous row.

Neckband

Join shoulder seams using a narrow backstitch. Using 3mm needles, slip the sts from the buttonhole band on to the needle and then knit up 29/30/31 sts up right front neck. Knit the 40/42/44 back neck sts on to the needle, knit up 29/30/31 sts down the other side of the neck and slip the buttonband sts on to the needle (118/122/126 sts). Work in double rib for 2cm. Over the next 2 rows form a buttonhole in line with those previously worked. Rib 2 more rows. Cast off in rib.

Making up

Join side and sleeve seams using a flat seam for the ribs and a narrow backstitch on the remainder (*see* Techniques, pages 17–18). Set sleeves in, taking care to distribute the fabric evenly around the armholes, easing any fullness to the top – i.e., the shoulder line. Join with a narrow backstitch. Attach buttons where marked.

TWO-TONE GUERNSEY

This man-sized cotton guernsey is simple and quick to knit. Worked in stocking stitch on large needles, two contrasting colours of double-knit cotton are knitted together to make the two-tone effect.

Materials
Melinda Coss DK cotton – 650gm of each of two contrasting colours.

Needles
One pair of 6½mm needles.

Tension
Using 6½mm needles and measured over st st, 13 sts and 18 rows = 10cm square.

Back and front
With 6½mm needles, cast on 98 sts, using one strand of each colour cotton.
Row 1: k2, *p2, k2, rep from * to end. Row 2: p2, *k2, p2, rep from * to end. Rep these 2 rows until work measures 4cm, ending with a 2nd row.
Cont in st st, inc 1 st each end of the first row (100 sts). Work straight until piece measures 69cm from beg.
Shape neck: cast off 30 sts at beg of next 2 rows. Work straight on remaining 40 sts for 6 rows. Cast off.

Sleeves
Using 6½mm needles, cast on 42 sts. Work in rib as for body for 4cm, ending with a 2nd row. Cont in st st, inc 1 st each end of the 3rd row and every following alt row until you have 82 sts. Cont without further shaping until work measures 42cm from beg. Cast off loosely.

Making up
Join shoulder seams with a flat seam (*see* Techniques, pages 18–19), sew sleeves to body, join sleeve and side seams. Press seams.

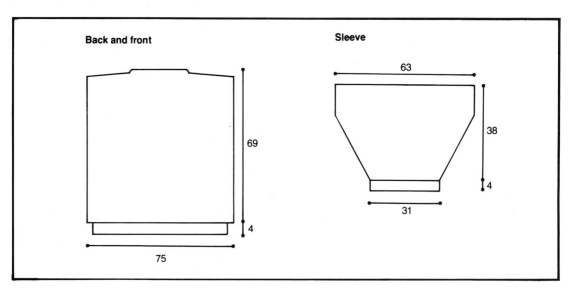

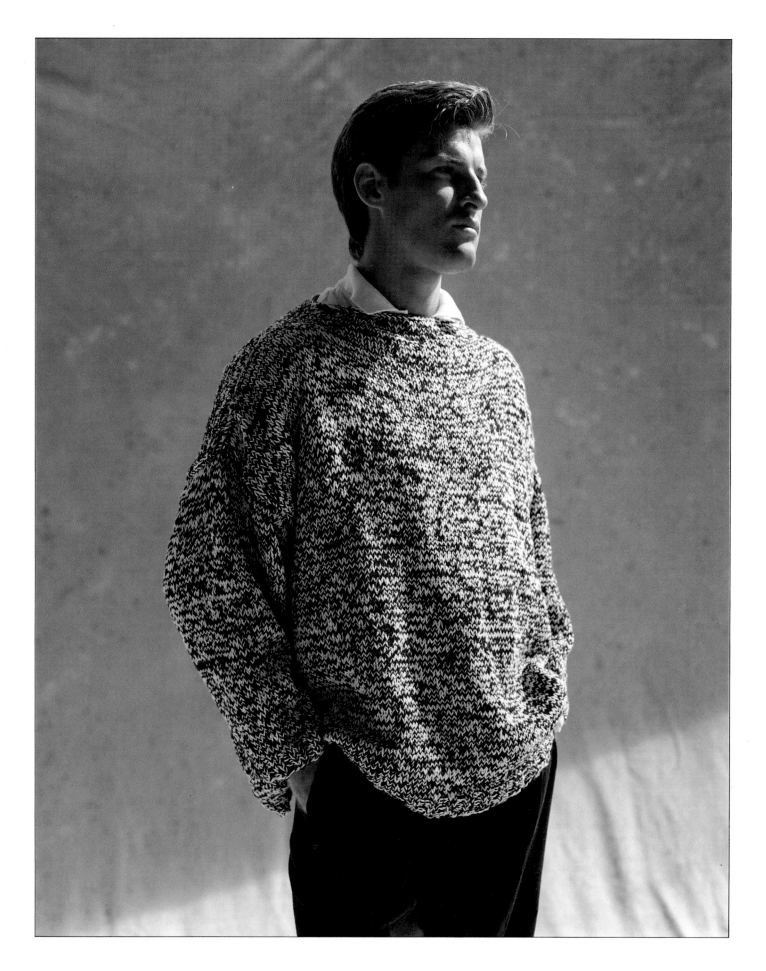

COTTON CORNFLOWER

A double-knit cotton jumper in two sizes (small/medium), featuring a pretty ribbed yoke and bobbles. The main body of the garment is worked in stocking stitch with a scattering of cornflowers.

Materials
Melinda Coss DK cotton – 650/700gm.

Needles
One pair of 3mm and one pair of 4mm needles; one 4mm circular needle.

Tension
Using 4mm needles and measured over st st, 20 sts and 24 rows = 10cm square.

Cornflower stitch pattern
Row 1: k1, yrn, k1, yrn, k1, in next st.
Row 2: purl new sts, wrapping yarn round needle twice for each st.
Row 3: with yarn at back, sl 4. Drop first elongated st off needle to front of work, slip the same 4 sts back to LH needle, pick up dropped st knitwise and slip it on to LH needle. Knit 2 tog through back loops (the elongated st and the next st). K3 with yarn in back, sl 3, dropping extra wraps. Drop last elongated st off needle to front. With yarn at back, sl 4, pick up dropped st on to LH needle. K3, k2 tog (the last of the 4 sts and the elongated st).
Row 4: with yarn in front, sl 3 off needle.
Row 5: with yarn in back, sl 3, drop next elongated st off needle to front, sl the same 3 sts back to LH needle, pick up dropped st knitwise and sl it on to LH needle. K2, k2 tog.
Row 6: purl.

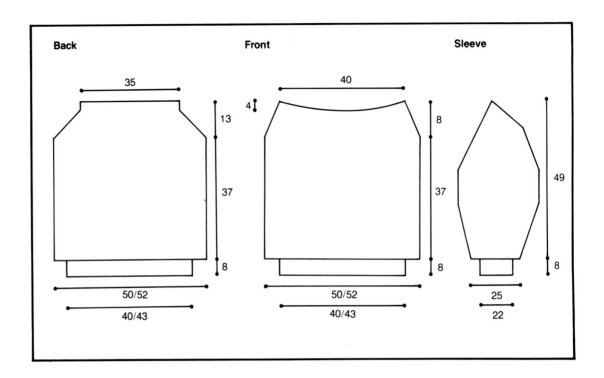

Bobble stitch pattern

K1, yrn, k1, yrn, k1, turn. P5, turn, k5, turn, p2 tog, p1, p2 tog, turn. Sl 1, k2 together, psso.

Back

Using 3mm needles, cast on 80/86 sts and work in k2, p2 rib for 20 rows, inc 20/24 sts evenly along last row of rib. Change to 4mm needles and commence in st st, working one cornflower stitch in each of the following positions.
Row 11: stitch 45/50. Row 17: stitch 80/85. Row 29: stitch 20/25. Row 35: stitch 60/65. Row 45: stitch 70/75. Row 55: stitch 40/45. Row 63: stitch 30/35. Row 73: stitch 75/80. Row 81: stitch 50/55. K straight to row 90*.
Size 1: dec 1 st each end of the next row and the 14 following alt rows (70 sts). Work 1 row.
Size 2: dec 1 st each end of every row 10 times, then dec 1 st each end of every alt row 10 times (70 sts).
At the same time, work cornflower stitches in these positions: row 95: stitch 32/35. Row 107: stitch 46. Work in st st to row 120. Cast off 35 sts. Mark with a coloured thread, cast off remaining 35 sts.

Front

Work as for back to *.
Dec 1 st each end of next and every following alt row (every following row) for 10 rows. At the same time work cornflower stitch on row 91, stitch 34/39. Cont to dec 1 st at both ends of next and every following alt row.
Shape neck: k36, place these sts on a stitch holder, cast off 16 sts. K to end of row, p 1 row. Cast off 8 sts. K to end of row, p 1 row, rep these last 2 rows 3 times more. Secure remaining stitch. Pick up sts on stitch holder, rep for 2nd side of neck, reversing all shapings.

Right sleeve

Using 3mm needles, cast on 44 sts and work 20 rows in k2, p2 rib, inc 6 sts evenly across last row. Change to 4mm needles and, commencing with a knit row, st st 2 rows. Inc 1 st each end of next and every following 4th row until you have 80 sts. At the same time work cornflower stitches in the following positions: Row 11: stitch 28. Row 21: stitch 16. Row 35: stitch 43. Row 47: stitch 33. Row 61: stitch 55. Row 75: stitch 25. Row 83: stitch 48. Work straight to row 90. Dec 1 st each end of next row and the 9 following alt rows, work 1 row (60 sts).*
Next row: cast off 11 sts, work to last 2 sts. Dec 1 st. Work 1 row. Rep these 2 rows until 1 st remains. Secure off.

Left sleeve

Work as for right sleeve to *, then cont as for right sleeve reversing shapings.

Shoulder and neckband

Using a 4mm circular needle and starting at thread marker, pick up 35 sts from centre back, leftwards, 55 sts across top of left sleeve, 80 sts across front, 55 sts across right sleeve and 35 sts across RS of back. Work 10 sts in moss st, then k3, p2 rib to last 10 sts. Work these in moss st. Next row: moss st 10. K2, p3 to last 10 sts. Moss st 10.

Make back opening: rep these 2 rows 3 times more, but work from edge to edge instead of completing the full circle. Work 1 more row in pattern.

Next row: moss st 8, k2 tog. Dec 48 sts along the row by dec on knit sts, so rib becomes k2, p2, ending with k2 tog. Moss st 8. *At the same time* MB on the first, then every following 4th dec st. Cont the moss st borders, work in k2, p2 rib for 9 rows. Next row: moss st 7, k2 tog, dec 48 sts along row by dec on purl stitches of rib so rib becomes k2, p1, ending with k2 tog. Moss st 7. *At the same time* make 13 bobbles in line with the previous bobbles. Cont the moss st border work in k2, p1 rib for 9 rows.

Next row: change to 3mm needles. Moss st 6, k2 tog, dec 48 sts along the row on k sts so rib becomes k1, p1, ending k2 tog, moss st 6. *At the same time* make 13 bobbles in line with previous bobbles. Work in k1, p1 rib for 7 rows.

Next row: moss st 5, k2 tog. Cont in rib, working bobbles in line with previous bobbles to last 7 sts. K2 tog, moss st 5. Work 1 more row, then cast off loosely knitwise over rib and ribwise over moss st.

Making up

Join underarm and side seams using a flat seam (*see* Techniques, pages 18–19). Make a small chain loop at the top of the moss st back opening and sew button in position on other side.

LACE-STITCH BLOUSE

A waisted, short-sleeved summer sweater with a Forties look. Worked from 4-ply mercerized cotton in a combination of two lace stitches, quoted in one size only.

Materials
Melinda Coss 4-ply mercerized cotton – 250gm; 4 glass buttons.

Needles
One pair of 2¾mm and one pair of 3¼mm needles.

Tension
Using 3¼mm needles and measured over st st, 30 sts = 10cm.

Back

Using 2¾mm needles, cast on 114 sts.
Row 1: *k1, p1, rep from * to end. Keep rep this row to form single rib for 9cm, ending with an RS row. Purl the next row, inc into every 10th st (125 sts). Change to 3¼mm needles and work in pattern.

Preparatory row:
K5, *(m1, k2 tog) 3 times, m1, k2, sl 1, k1, psso, k10. Rep from * to end. Next row: purl.
Row 1: k5, *m1, k1, (m1, k2 tog) 3 times, m1, k2, sl 1, k1, psso, k5, k2 tog, k2. Rep from * to end.
Row 2 and each alt row: purl.
Row 3: k5, *m1, k3, (m1, k2 tog) 3 times, m1, k2, sl 1, k1, psso, k3, k2 tog, k2. Rep from * to end.
Row 5: k5, *m1, k5, (m1, k2 tog) 3 times, m1, k2, sl 1, k1, psso, k1, k2 tog, k2. Rep from * to end.
Row 7: k5, *m1, k7, (m1, k2 tog) 3 times, m1, k2, sl 1, k2 tog, psso, k2. Rep from * to end.

Row 9: k2, *sl 1, k1, psso, k5, k2 tog, k2, (m1, k2 tog) 3 times, m1, k1, m1, k2. Rep from * to last 3 sts, k3.
Row 11: k2, *sl 1, k1, psso, k3, k2 tog, k2, (m1, k2 tog) 3 times, m1, k3, m1, k2. Rep from * to last 3 sts, k3.
Row 13: k2, *sl 1, k1, psso, k1, k2 tog, k2, (m1, k2 tog) 3 times, m1, k5, m1, k2. Rep from * to last 3 sts, k3.
Row 15: k2, *sl 1, k2 tog, psso, k2, (m1, k2 tog) 3 times, m1, k7, m1, k2. Rep from * to last 3 sts, k3.
Row 16: purl.
These 16 rows form the pattern. Rep rows 1–16 inclusive 5 times in all, then rep rows 1–8 inclusive once more.

Shape armholes and commence yoke:
Row 1: cast off 10 sts, (m1, k2 tog) 24 times, m1, k1, m1, k2, sl 1, k1, psso, k5, k2 tog, k2, (m1, k2 tog) 25 times, k2.
Row 2: cast off 11 sts, p to end.
Row 3: k2 tog, k1, (m1, k2 tog) 23 times, m1, k3, m1, k2, sl 1, k1, psso, k3, k2 tog, k2, (m1, k2 tog) 19 times, k1, k2 tog.
Row 4: p2 tog, p to last 2 sts, p2 tog.
Row 5: k2 tog, k1, (m1, k2 tog) 22 times, m1, k5, m1, k2, sl 1, k1, psso, k1, k2 tog, k2, (m1, k2 tog) 18 times, k1, k2 tog.
Row 6: as row 4.
Row 7: k2 tog, k1, (m1, k2 tog) 21 times, m1, k7, m1, k2, sl 1, k2 tog, psso, k2, (m1, k2 tog) 17 times, k1, k2 tog.
Row 8: as row 4 (92 sts remain).
Row 9: k2, (m1, k2 tog) 21 times, m1, k2, sl 1, k1, psso, k5, k2 tog, k2, m1, k1, (m1, k2 tog) 16 times, k2.
Row 10 and each alt row: purl.
Row 11: k3, (m1, k2 tog) 21 times, m1, k2, sl 1, k1, psso, k3, k2 tog, k2, m1, k3, (m1, k2 tog) 16 times, k1.
Row 13: k2, k2 tog, (m1, k2 tog) 21 times, m1,

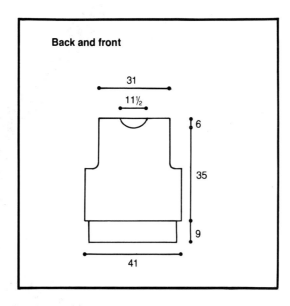

Back and front

k2, sl 1, k1, psso, k1, k2 tog, k2, m1, k5, (m1, k2 tog) 15 times, m1, k2.
Row 15: k2, (m1, k2 tog) 22 times, m1, k2, sl 1, k2 tog, psso, k2, m1, k7, (m1, k2 tog) 15 times, k2.
Row 17: k2, k2 tog, (m1, k2 tog) 21 times, m1, k1, m1, k2, sl 1, k1, psso, k5, k2 tog, k2, (m1, k2 tog) 15 times, m1, k2.
Row 19: k3, (m1, k2 tog) 21 times, m1, k3, m1, k2, sl 1, k1, psso, k3, k2 tog, k2, (m1, k2 tog) 16 times, k1.
Row 21: k1, k2 tog, (m1, k2 tog) 21 times, m1, k5, m1, k2, sl 1, k1, psso, k1, k2 tog, k2, (m1, k2 tog) 16 times, m1, k1.
Row 23: k2, k2 tog, (m1, k2 tog) 20 times, m1, k7, m1, k2, sl 1, k2 tog, psso, k2, (m1, k2 tog) 16 times, m1, k2.
Row 24: purl.
Rep rows 9–24 inclusive until work measures 52cm, ending with a WS row. Cast off 27 sts, pattern 38 sts and put these on a holder, pattern to end. Pattern 2 more rows on remaining sts before casting off very neatly.

Front
Work as for back until the work measures 44cm, ending with a WS row.
Shape neck: pattern 60 sts, place these sts on a holder and pattern to end. Cont with this set of sts, dec 1 st at neck edge until 27 sts remain. Now work straight until front matches back. Return to held sts, leave the centre 28 sts on the holder, join yarn in at neck edge and shape the other side of the neck as for the first side. Now work straight until this side is the same length as the other and then work for 3 more rows before

casting off very neatly (this edge will be the buttoning edge).

Sleeves
Using 2¾mm needles, cast on 86 sts and work in single rib for 4cm, ending with an RS row. Purl the next row, inc into every 14th st (92 sts). Change to 3¼mm needles and work as for yoke. Rep rows 9–24 inclusive, twice.
Shape sleeve head:
Row 1: k2 tog, (m1, k2 tog) 19 times, m1, k2, sl 1, k1, psso, k5, k2 tog, k2, m1, k1, (m1, k2 tog) 18 times, k2 tog.
Row 2 and each alt row: p2 tog, p to last 2 sts, p2 tog.
Row 3: k2 tog, k1, (m1, k2 tog) 18 times, m1, k2, sl 1, k1, psso, k3, k2 tog, k2, m1, k3, (m1, k2 tog) 16 times, k1, k2 tog.
Row 5: k2 tog, k2 tog, (m1, k2 tog) 17 times, m1, k2, sl 1, k1, psso, k1, k2 tog, k2, m1, k5, (m1, k2 tog) 15 times, m1, k2 tog.

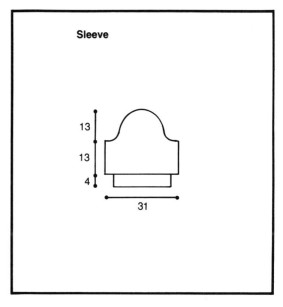

Sleeve

Row 7: k2 tog, k2, (m1, k2 tog) 16 times, m1, k2, sl 1, k2 tog, psso, k2, m1, k7, (m1, k2 tog) 14 times, k2 tog.
Row 9: k2 tog, k2 tog, (m1, k2 tog) 15 times, m1, k1, m1, k2, sl 1, k1, psso, k5, k2 tog, k2, (m1, k2 tog) 13 times, k1, k2 tog.
Row 11: k2 tog, k1, (m1, k2 tog) 14 times, m1, k3, m1, k2, sl 1, k1, psso, k3, k2 tog, k2, (m1, k2 tog) 12 times, k1, k2 tog.
Row 12: as row 2.
Row 13: k2 tog, (m1, k2 tog) 14 times, m1, k5, m1, k2, sl 1, k1, psso, k1, k2 tog, k2 (m1, k2 tog) 12 times, k2 tog.
Row 14: as row 2.
Row 15: k2 tog, k2 tog, (m1, k2 tog) 11 times,

m1, k7, m1, k2, sl 1, k2 tog, psso, k2, (m1, k2 tog) 11 times, m1, k2 tog.

Row 16: as row 2 (60 sts). Now rep rows 9–24 inclusive as given for the yoke, twice, but with the following alterations: for 20 times work 12 times; 19 times, 11 times; 18 times, 10 times; 17 times, 9 times,

Now rep rows 1–12 inclusive for the sleeve shaping, once with the following alterations: for 16 times work 8 times; 15 times, 7 times; 14 times, 6 times; 13 times, 5 times; 12 times, 4 times; 11 times, 3 times.

Cast off remaining sts.

Neckband

Join the right shoulder seam with a flat seam. Using 2¾mm needles and with RS facing, knit up 24 sts down left front, knit across the held sts at front, knit up 22 sts up other side and then knit across the back neck sts (106 sts). Purl the first (WS) row, then work in single rib for 6 rows.

Work buttonhole: next row: rib 2, cast off 2, rib to end. Rib the next row, casting on 2 sts above those cast off on previous row. Cast off loosely in rib.

Making up

Join side and sleeve seams with a flat seam over rib, a narrow backstitch over pattern (*see* Techniques, pages 17–18). Join the right shoulder seam with a narrow backstitch. Overlap the left shoulder edge over the back edge by the extra rows. Set sleeves in with a very narrow backstitch, easing any fullness to the top.

Attach buttons.

LACE SWEATER

A delicate lacy one-size sweater with deep twisted ribbing at cuff and hip and picot edging around the neck. The stitch used is not a complicated one, the repeat being over 14 stitches and 12 rows, but it does involve working into the backs of stitches, which slows down the working and often tightens tension. Take extra care when checking your tension.

Materials
Melinda Coss 4-ply mercerized cotton – 500gm.

Needles
One pair each of 2¼mm, 2¾mm and 3¼mm needles.

Tension
Using 3¼mm needles, cast on 31 sts and work, in pattern, for 24 rows. Cast off and lightly press. The sample should measure 10cm across, not counting the edge sts, and 6cm down, not counting the cast-on and cast-off rows.

Pattern
Row 1 (RS): k1, *p2, k1, sl 1, psso, k3-b, yo, k1-b, yo, k3-b, k2 tog, p1. Rep from *, p1, k1.
Row 2: k2, *k1, p4-b, p1, p1-b, p1, p4-b, k2. Rep from *, k1.
Row 3: k1, *p2, k1, sl 1, psso, k2-b, yo, k3-b, yo, k2-b, k2 tog, p1. Rep from *, p1, k1.
Row 4: k2, *k1, (p3-b, p1) twice, p3-b, k2. Rep from *, k1.
Row 5: k1, *p2, k1, sl 1, psso, k1-b, yo, k5-b, yo, k1-b, k2 tog, p1. Rep from *, p1, k1.
Row 6: k2, *k1, p2-b, p1, p5-b, p1, p2-b, k2. Rep from *, k1.
Row 7: k1, *k1-b, yo, k3-b, k2 tog, p3, k1, sl 1,

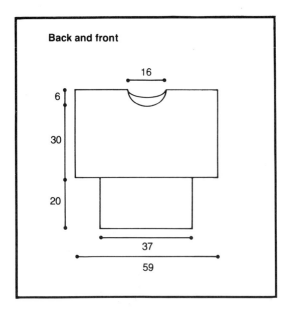

Back and front

16

6

30

20

37

59

psso, k3-b, yo. Rep from *, k1-b, kl.
Row 8: k1, p1-b, *p1, p4-b, k3, p4-b, p1, p1-b. Rep from *, k1.
Row 9: k1, *k2-b, yo, k2-b, k2 tog, p3, k1, sl 1, psso, k2-b, yo, k1-b. Rep from *, k1-b, k1.
Row 10: k1, p1-b, *p1-b, p1, p3-b, k3, p3-b, p1, p2-b. Rep from *, k1.
Row 11: k1, *k3-b, yo, k1-b, k2 tog, p3, k1, sl 1, psso, k1-b, yo, k2-b. Rep from *, k1-b, k1.
Row 12: k1, p1-b, *p2-b, p1, p2-b, k3, p2-b, p1, p3-b. Rep from *, k1.
These 12 rows form the pattern.

Back

Using 2¾mm needles, cast on 162 sts.
Row 1 (WS): * k1, p1, rep from * to end.
Row 2: * k1 into back of st, p1, rep from * to end. Keep rep this row to form twisted rib for 8cm. Change to 2¼mm needles and work until it measures 11cm. Return to 2¾mm needles and cont until work measures 20cm, ending with an RS row. Purl the next row, inc into every 18th st (171 sts). Change to 3¼mm needles and cont in pattern. When 11½ patterns have been worked, **shape neck**: row 7 of pattern: work 64 sts, cast off 43 sts, work to end. Cont with this set of sts, leaving others on a holder. Dec 1 st at neck edge on every row until 62 sts remain, keeping in pattern. Complete the pattern. Leave sts on a holder. Return to other set of sts, join yarn in at neck edge and shape to match the other side. Leave sts on a spare needle.

Front

Work as for back until 10 pattern repeats have been worked.
Shape neck: row 1 of pattern: work 71 sts, cast off 29 sts, work to end. Cont with this set of sts, leaving others on a holder. Dec 1 st at neck edge on every RS row, keeping in pattern, until 62 sts remain. Cont straight until the pattern is complete. Leave sts on a holder. Return to other set of sts, join yarn in at neck edge and shape to match the first side. Leave sts on a spare needle.

Sleeves

Using 2¾mm needles, cast on 62 sts and work in twisted rib for 12cm, ending with a WS row. Next row: knit, inc into every 3rd st (82 sts). Next row: p1, *p1, inc into next st, rep from * 8 more times, p3. **Inc into next st, p2, rep from ** to last 20 sts, inc into every alt st to end (115 sts). Change to 3¼mm needles and cont in pattern, inc 1 st each end of every 8th row, keeping the new sts in reverse st st (purl on RS rows, knit on WS rows), until you have 143 sts. Work straight until the pattern is complete. Cast off loosely.

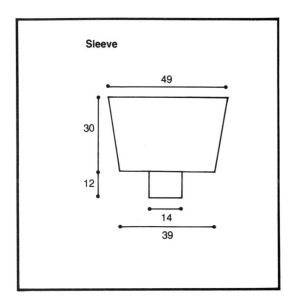

Sleeve

49

30

12

14

39

Neckband

Knit right shoulder seam together fairly firmly (this stitch has a tendency to stretch, which is undesirable on a shoulder seam; *see* Techniques, pages 19–20). Using a 2¼mm needle and with RS of work facing, knit up 19 sts down the left front, 26 sts across the front, 18 sts up the other side and 54 sts around the back (117 sts), taking great care to do so evenly so that the eventual picot edge will sit properly. Using 3mm needles, work 3 rows in st st. Next row (RS): *k2 tog, yo, keep rep from * to last st, k1. Change to 2¾mm needles and work 3 more rows in st st. Leave sts on a thread.

Making up

Knit the left shoulder seam tog and join the edges of the picot border with a flat seam (*see* Techniques, pages 18–19). Turn neck edging in so that the picot points stand up and slip st down to the line below the original knitted up sts. Take care not to twist the picot as you do this. Remove the thread after all sts are secured.

Open out the body and pin the sleeves into position, taking care not to bunch them (slightly stretch rather than bunch them). Join with a narrow backstitch. Join side and sleeve seams with a flat seam over ribs, backstitch over pattern.

DIAMOND-FRONT CARDIGAN

The diamond pattern on the front of this one-size cardigan is worked using the intarsia method (*see* Techniques, pages 13–14). It may be transformed into an argyll-type design simply by adding some (optional) Swiss darning after completion. The long line and drop shoulder make it an easy-to-wear style.

Materials
Melinda Coss 4-ply matt cotton – lemon: 450gm; white: 50gm; blue: 50gm; 7 buttons.

Needles
One pair of 2¾mm and one pair of 3¼mm needles.

Tension
Using 3¼mm needles and measured over st st, 28 sts and 34 rows = 10cm square.

Right front
Using 2¾mm needles and lemon, cast on 76 sts.
Row 1: *k2, p2, rep from * to end. Keep rep this row to form double rib for 6cm, ending with an RS row. Purl the next row, inc 4 sts evenly across it. Now change to 3¼mm needles and work pattern from graph until work measures 42cm, ending with a WS row.
Shape neck (meanwhile cont in pattern): dec 1 st at neck edge on next and every following 5th row until 58 sts remain. Work straight until the 36th row of the 4th pattern repeat has been worked. Leave sts on a spare needle.

Left front
Work as for right front, reversing out neck shaping.

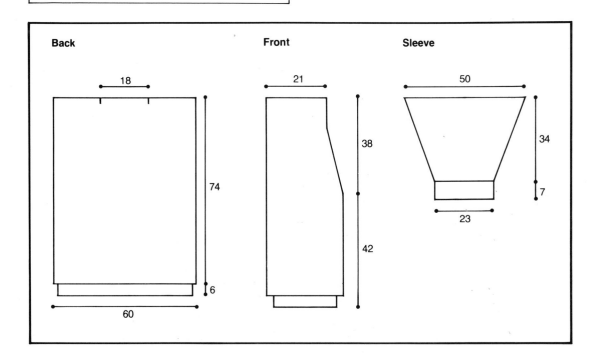

Back

Using 2¾mm needles and lemon, cast on 160 sts and work in double rib for 6cm, ending with an RS row. Purl the next row, inc 8 sts evenly across it. Change to 3¼mm needles and cont working in st st until the back matches the front. Leave sts on a spare needle.

Incorporate the graph below into the right and left fronts of the cardigan, adding "argyll" criss-crosses by Swiss darning if required.

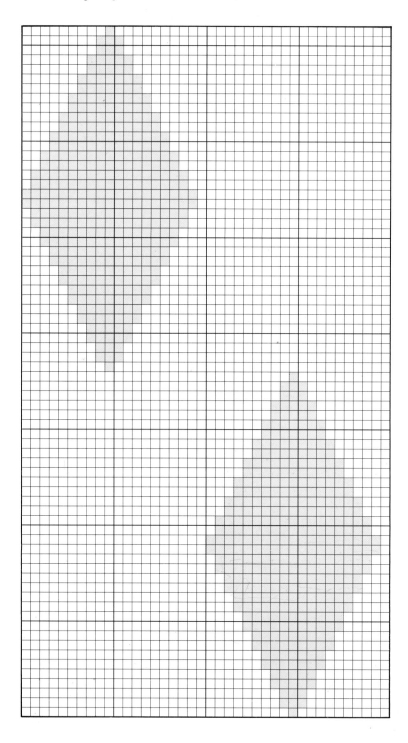

Sleeves

Using 2¾mm needles and lemon, cast on 52 sts and work in double rib for 7cm, ending with a WS row. Knit the next row, inc into every 4th st (65 sts). Change to 3¼mm needles and cont in st st, inc 1 st each end of every 3rd row until you have 141 sts. Work straight until sleeve measures 41cm. Cast off loosely.

Bands

Using 2¾mm needles and lemon, cast on 214 sts and work in double rib for 4 rows.
Row 5 (RS): rib 4, *cast off 2, rib 14, rep from * 6 times more, rib to end.
Row 7: rib, casting on 2 sts above those cast off on previous row.
Rib 2 more rows. Cast off knitwise, using a larger needle to obtain an edge with the correct tension.
Work another band to match but omit the buttonholes.

Making up

Knit 1 shoulder seam tog (*see* Techniques, pages 19–20), cast off the 52 back neck sts, knit second shoulder seam tog.
Open work out and pin the sleeves in position, taking care not to bunch them (slightly stretch them to avoid bunching), and distributing them equally either side of the shoulder seam. Join with a narrow backstitch.
Join side and sleeve seams with a flat seam over ribs, a backstitch over the st st. Join the bands with flat seam and centre this at the back neck. Pin and attach with a flat seam, distributing them evenly around the fronts. Attach buttons to correspond to buttonholes.
If "argyll" criss-crosses are required, these should be worked by Swiss darning (*see* Techniques, pages 20–21). The cardigan that was photographed has had them worked at random, in lemon, but they may be placed according to taste.

ROLL-NECK DEEP RAGLAN SWEATER

A cool classic of ample proportions, worked in 4-ply mercerized cotton. The deep raglan is accentuated with extra shaping and shoulder pads. The sizes quoted are women's medium/large.

Materials
Melinda Coss 4-ply mercerized cotton – both sizes: 600gm; one pair of raglan shoulder pads.

Needles
One pair each of 2¾mm, 3mm and 3¼mm needles.

Tension
Using 3¼mm needles and measured over st st, 30 sts = 10cm.

Back
Using 2¾mm needles, cast on 164/168 sts. Row 1: k2, p2, to end. Keep rep this row to form double rib for 6cm, ending with a WS row. Knit the next row, inc into every 16th/14th st (174/180 sts). Now change to 3¼mm needles and cont working in st st until the work measures 42/44cm, ending with a WS row. **Shape raglan**: cast off 4 sts at beg of next 2 rows.
Row 3: k3, sl 1, k1, psso, k to last 5 sts, k2 tog, k3.
Row 4: p3, p2 tog, p to last 4 sts, sl the last worked st back on to the LH needle, lift the 2nd st on the LH needle over it, return it to the RH needle, p3. Cont thus, dec 1 st each end of every row until 134/152 sts remain. Now cont to dec on RS rows only until 42/46 sts remain. Work 1 row straight and then cast off.

Front
As for back until 64/66 sts remain. **Shape neck** (meanwhile cont to shape raglan as before): next row: work 22/21 sts, cast off 20/24 sts, work to end. Cont with this last set of sts, leaving the first on a holder. Dec 1 st at neck edge on every row until 9 decs have been made. Now work this edge straight, meanwhile cont to shape raglan until 2 sts remain. Work 1 row straight. Work these 2 sts tog and secure. Return to the held sts and work the other side of the neck to match.

Sleeves
Using 2¾mm needles, cast on 56/60 sts and work in double rib for 6cm, ending with a WS row. Knit the next row, inc 6 sts evenly across it (62/66 sts). Now change to 3¼mm needles and cont in st st, inc 1 st each end of next and every following 3rd row until you have 144/150 sts. Work straight until the sleeve is 41/44cm from beg, measured on the straight, ending with a WS row. **Shape raglan**: cast off 4/5 sts at beg of next 2 rows. Now dec 1 st each end of every knit row, as on body, until 44/34 sts remain. Now dec 1 st each end of every row until 12/14 sts remain. Work 1 row straight and then cast off.

Collar
First join 3 raglan seams with a very narrow backstitch, leaving the left back one open. Using 3mm needles and with RS of work facing, knit up 140/144 sts around the neck, taking care to space the sts evenly. Purl the first row (WS) and then cont in double rib until the collar measures 16cm. Cast off loosely in rib.

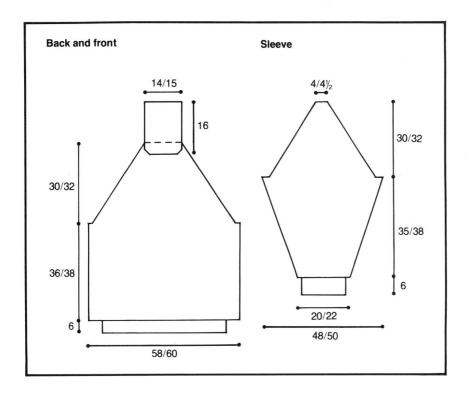

Back and front

14/15

16

30/32

36/38

6

58/60

Sleeve

4/4½

30/32

35/38

6

20/22

48/50

Making up

Join the collar seam from the top downwards with a flat seam (*see* Techniques, pages 18–19). Cont this seam down the remaining raglan with a backstitch. Join side and sleeve seams with a flat seam over the ribs and a backstitch over the st st.

FISHERMAN'S CABLE SWEATER

A chunky man-sized fisherman's sweater worked in double-knit cotton and featuring an intricate cable stitch, moss stitch borders and bold stripes. A challenging project for the more experienced knitter.

Materials
Melinda Coss DK cotton – navy: 1,000gm; white: 250gm.

Needles
One pair of 4mm needles; one 4mm circular needle.

Tension
Using 4mm needles and measured over pattern, 27 sts and 24 rows = 10cm square.

Abbreviations
t3b: slip 1 st on to a cable needle and hold at back; k2 from LH needle, p1 from cable needle. **t3f**: slip 2 sts on to a cable needle and hold at front; p1 from LH needle, k2 from cable needle. **c4f**: slip 2 sts on to a cable needle and hold at front; k2 from LH needle, k2 from cable needle. **c4b**: slip 2 sts on to a cable needle and hold at back; k2 from LH needle, k2 from cable needle.

Front
Using 4mm needles and navy, cast on 176 sts. K2, p2 rib for 2½cm, ending with a WS row.
Next row: *k2 in navy, p2 in white, rep from * to end. Next row: *k2 in white, p2 in navy, rep from * to end. Rep these 2 rows twice more.

Pattern
Row 1: k 16 sts, in white. Change to navy, p1. *T3f, t3b, p2; rep from * 16 times more.

T3f, t3b, p1. Change to white, k16.
Row 2: p16 in white. Change to navy, k2. *P4, k4; rep from * 16 times more. P4, k2. Change to white, k16.
Row 3: in navy, k16, p2. *C4f, p4; rep from * 16 times more. C4f, p2, k16.
Row 4: in navy, p16, k2. *P4, k4; rep from *

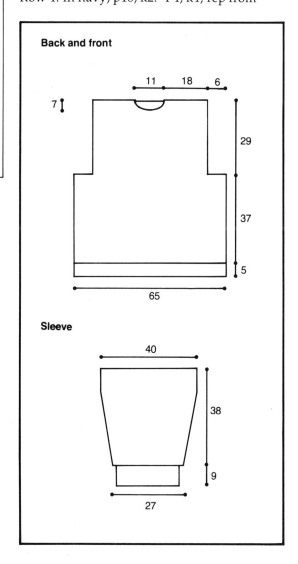

Back and front

Sleeve

16 times more. P4, k2, p16.

Row 5: k16 in white. Change to navy, p1. *T3b, t3f, p2; rep from * 16 times more. T3b, t3f, p1. Change to white, k16.

Row 6: p16 in white. Change to navy, k1. *P2, k2; rep from * to last 19 sts. P2, k1. Change to white, p16.

Keeping the first and last 16 sts in the 2-row striped sequence, rep the last 6 rows.

Row 13: k16 in white. Change to navy. *T3b, p2, t3f; rep from * 17 times more. Change to white, k16.

Row 14: p16 in white. Change to navy, p2. *K4, p4; rep from * 16 times more. K4, p2. Change to white, p16.

Row 15: k18 in navy. *P4, c4b; rep from * 16 times more. P4, k18.

Row 16: p18 in navy. *K4, p4; rep from * 16 times more. K4, p18.

Row 17: k16 in white. Change to navy, k2, p3. *T3b, t3f, p2; rep from * 16 times more. P1, k2. Change to white, k16.

Row 18: p16 in white. Change to navy, p2, k3. *P2, k2; rep from * to last 19 sts. K1, p2. Change to white, p16.

Row 19: k18 in navy, p3. *T3f, t3b, p2; rep from * 16 times more. P1, k18.

Row 20: work as for row 16.

Row 21: k16 in white. Change to navy, k2. *P4, c4b; rep from * 16 times more. P4, k2. Change to white, k16.

Row 22: work as for row 14.

Row 23: k16 in navy. *T3f, p2, t3b; rep from * 17 times more. K16.

Row 24: p16 in navy, k1. *P2, k2; rep from * to last 19 sts. P2, k1, p16.

These 24 rows make up your pattern. Rep them 3 times, then rep the first 17 rows.

Shape armhole: cast off 16 sts at beg of next 2 rows. Cont in pattern without further shaping until work measures 59cm from beg, ending with a WS row.

Shape neck: commencing in navy, cont working in st st stripe sequence of 2 rows in navy, 2 rows in white, shaping the neck at the same time thus: k57 in navy. Slip these sts on to a spare needle, cast off 30 sts, k to end. Working on this last set of sts only, purl back to neck edge. Change to white and cont stripe sequence. Dec 1 st at neck edge on next 9 rows. Work 5 rows and cast off. Rep shaping for other side of neck.

Back

Work as for front until beg of neck shaping. Do not shape neck but cont in st st stripe sequence until back length matches front. Cast off.

Sleeves

Using 4mm needles and navy, cast on 56 sts.

Row 1: k2 in navy, p2 in white, rep to end.

Row 2: k2 in white, p2 in navy, rep to end.

Rep these 2 rows until work measures 7cm. Cont in 2 × 2 rib using navy only for another 2cm, inc 18 sts evenly along last row of rib ending on WS (74 sts).

Row 1: with navy, and starting with a knit st, moss st 13. *P1, t3f, t3b, p2, t3f, t3b, p1.* K16, rep from * to *, moss st 13.

Row 2: inc 1, moss st 13. *K2, p4, k4, p4, k2.* P16, rep from * to *, moss st 13, inc 1.

Row 3: moss st 14. *P2, c4f, p4, c4f, p2.* Join in white, k16. Change to navy. Rep from * to *, moss st 14.

Row 4: inc 1, moss st 14. *K2 , p4, k4, p4, k2.* P16 in white, change to navy. Rep from * to *, moss st 14, inc 1.

Row 5: moss st 15. *P1, t3b, t3f, p2, t3b, t3f, p1, k16.* Rep from * to *, moss st 15.

Row 6: inc 1, moss st 15. *K1, p2, k2, p2, k2, p2, k2, p2, k1.* P16, rep from * to *, moss st 15, inc 1.

Cont to inc 1 st each side on every alt row and keeping the central st st panel of 16 sts in 2-row st st stripe sequence, rep the last 6 rows.

Row 13: moss st 18. *T3b, p2, t3f, t3b, p2, t3f.* K16 in navy, rep from * to *, moss st 18.

Row 14: inc 1, moss st 18. *P2, k4, p4, k4, p2.* P16 in navy. Rep from * to *, moss st 18, inc 1.

Row 15: moss st 19. *P4, cb4, p8.* K16 in white. Rep from * to *, moss st 19.

Row 16: inc 1, moss st 19. *P2, k4, p4, k4, p2.* P16 in white. Rep from * to *, moss st 19, inc 1.

Row 17: moss st 20. *K2, p3, t3b, t3f, p3, k2.* K16 in navy. Rep from * to *, moss st 20.

Row 18: inc 1, moss st 20. *P2, k3, p2, k2, p2, k3, p2.* P16 in navy. Rep from * to *. Moss st 20, inc 1.

Row 19: moss st 21. *K2, p3, t3f, t3b, p3, k2.* K16 in white. Rep from * to *, moss st 21.

Row 20: inc 1, moss st 21. *P2, k4, p4, k4, p2.* P16 in white. Rep from * to *, moss st 21, inc 1.

Row 21: moss st 22. *K2, p4, c4b, p4, k2.* K16 in navy. Rep from * to *. Moss st 22.

Row 22: moss st 22. *P2, k4, p4, k4, p2.* P16 in navy. Rep from * to *. Moss st 22.

Row 23: inc 1, moss st 22. *T3f, p2, t3b, t3f,

p2, t3b.* K16 in white. Rep from * to *. Moss st 22, inc 1.

Row 24: moss st 23. *K1, p2, k2, p2, k2, p2, k2, p2, k1.* P16 in white. Rep from * to *. Moss st 23.

Rep rows 1 and 2 in navy only, starting row 1. Inc 1, moss st 23 and finishing moss st 23, inc 1. Start and finish row 2 with moss st 24. You have now finished your arm-patch. Cont following the cable pattern row by row but rep each * to * instruction one extra time over the position where you have been working the central patch – i.e., you will have 3 repeats of each cable instruction along each row instead of 2. Cont to increase the moss st panels either side on the next alt row and every following 3rd row until you have 108 sts. Cont without further shaping in moss st only until work measures 47cm from cast on. Cast off loosely.

Collar

Join shoulder seams taking care to match stripes. With 4mm circular needle and navy, pick up 104 sts evenly around the neck. Work in rounds of k2 in navy, p2 in white, rib for 10cm. Cast off in navy.

Making up

Set sleeves into armhole shaping and join using a flat seam (*see* Techniques, pages 18–19). Join side seams, taking care to match stripes by using appropriate coloured yarn for each stripe. Join sleeve seams.

BUTTERFLY

A dolman evening sweater with dramatic V-shape front and back, which uses a combination of silky mercerized and flat cotton. The sweater is knitted using the intarsia method.

Materials

Melinda Coss 6-ply mercerized cotton – black: 400gm; red: 50gm; gold: 100gm; lemon: 100gm; coffee: 100gm; khaki: 100gm. Melinda Coss 6-ply flat cotton – coral: 50gm.

Needles

One pair of 2¾mm and one pair of 3¼mm needles.

Tension

Using 3¼mm needles and measured over st st, 32 rows and 26 sts = 10cm square.

Front and back

Using 2¾mm needles and black, cast on 112 sts. K1, p1 rib for 10cm. Change to 3¼mm needles and commence following the graph in st st working first 20 rows in black only as indicated. When 28 rows have been completed, commence shaping by increasing 1 st each end of the next row and the 6 following 3rd rows. Then inc each end of every alt row 24 times (174 sts). Cast on 12 sts at beg of next row, then work 84 sts. Cast off 6 sts for neck, slip these first 96 sts on to a spare needle, then cont for the other side only by working to the end of the row. Inc 12 sts at beg of next row and the 2 following alt rows. *At the same time* **shape neck**: dec 1 st at neck edge on every 3rd row 21 times (111 sts). Cont without further shaping for 8 rows, leave sts on a spare needle. Return to the other side of the work, rejoin yarn at inner edge, work to match.

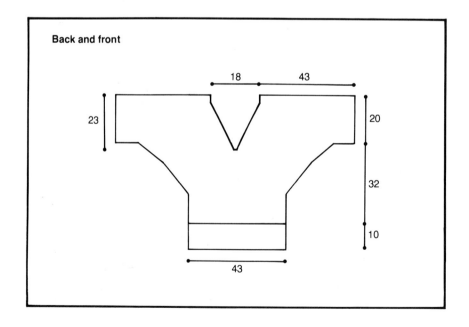

Back and front

Incorporate this graph into
both the front and the back of
the sweater.

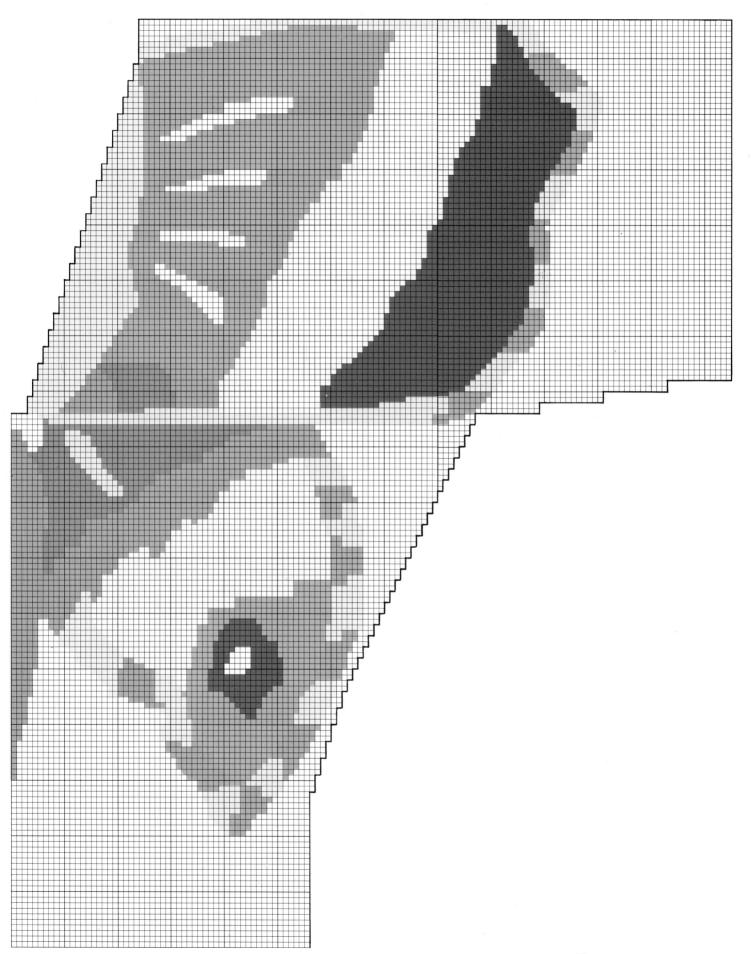

When the front and back are completed, graft shoulder/top sleeve seams together using the colours in the design. Secure all ends.

Cuffs

With 2¾mm needles and RS of work facing, pick up and knit every other stitch along cuff edge (65 sts).

Row 1: k1, p2 tog. *K1, p1, k1; rep from * to last 3 sts. P1, k1, p1. Cont in k1, p1 rib, dec 1 st each end of the following 5th row and every following 8th row until you have 44 sts. Work 8 rows in rib without shaping. Cast off using a larger needle.

Neckbands

With RS of work facing and commencing at the base of the RS of the "V" front neck shaping, use 2¾mm needles and pick up and knit 122 sts in black, ending at the base of the "V" back neck shaping. K1, p1 rib for 10 rows. Cast off. Rep for other side of neck shaping.

Making up

Stitch the base of the right neck band along the bottom of the "V" front and back, then stitch the base of the left neckband directly over the right neckband front and back. Join side seams using a flat seam (*see* Techniques, pages 18–19).

JUNGLE-PRINT JACKET WITH TUBE CAMISOLE

A tropical two-piece for warm summer nights and lazy days. Worked in double-knit cotton using the intarsia method, this boxy jacket and tube top are ideal holiday wear.

JACKET

Right front
Using 3mm needles and red, cast on 54 sts. K1, p1 rib for 6 rows. Change to 3¾mm needles and blue. Commence following the graph in st st starting with a knit row. Work straight until you have completed 71 rows. Shape armhole by casting off 7 sts at beg of

Materials
Melinda Coss DK cotton – blue: 450gm; red: 150gm; green: 50gm; yellow: 50gm; black: 50gm; orange: 50gm; 8 buttons.

Needles
One pair each of 3mm, 3¾mm and 4mm needles.

Tension
Using 3¾mm needles and measured over st st, 20 sts and 28 rows = 10cm square.

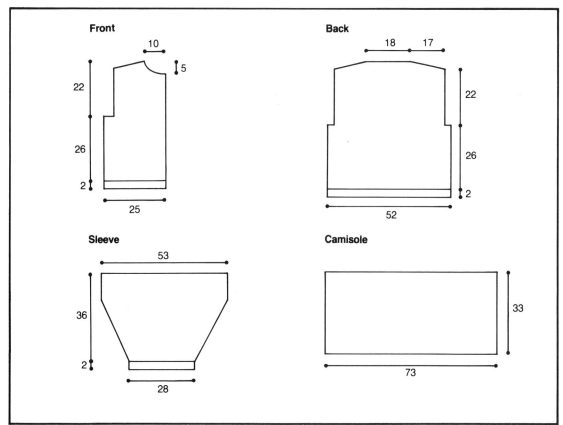

Front

10

5

22

26

2

25

Back

18   17

22

26

2

52

Sleeve

53

36

2

28

Camisole

33

73

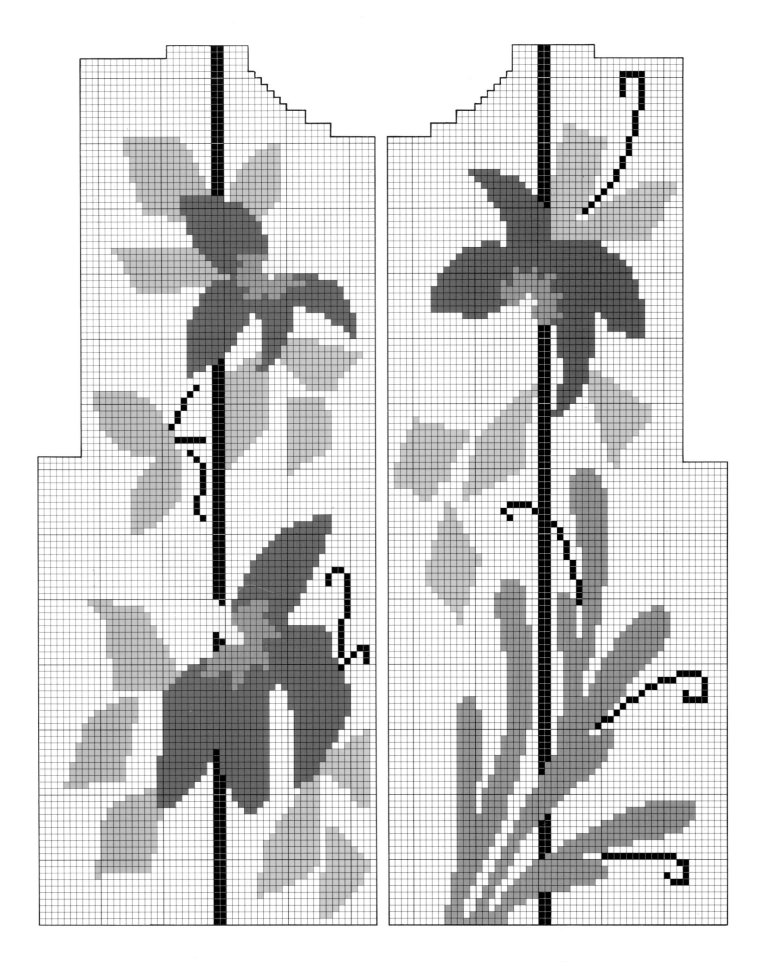

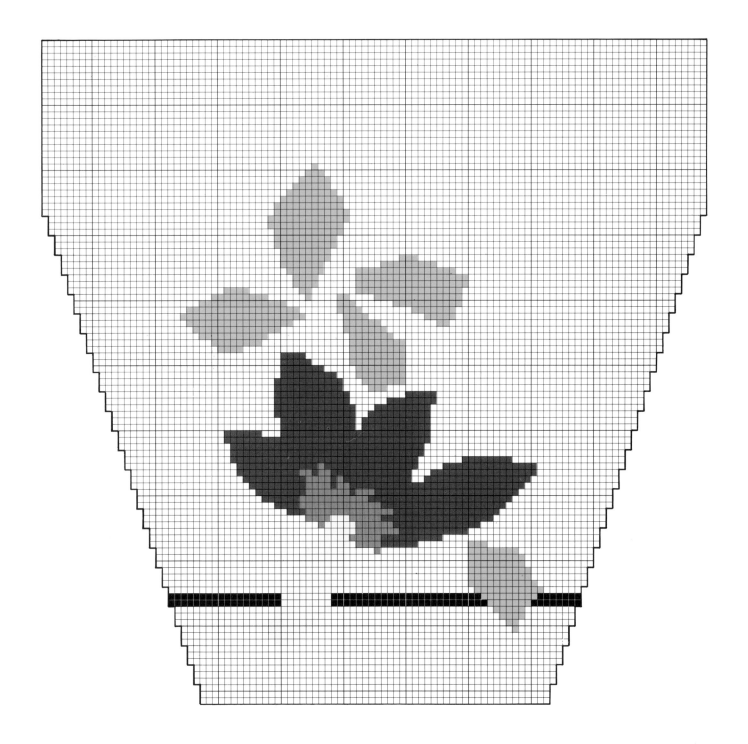

Incorporate the graphs on page 67 into the left and right fronts of the jacket and the graph above into the left sleeve.

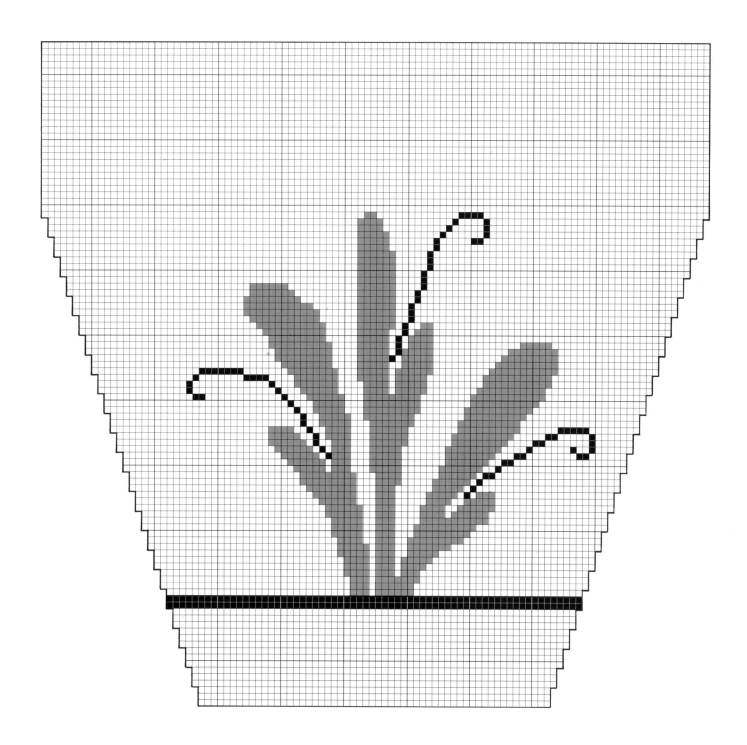

Incorporate this graph into
the right sleeve of the jacket.

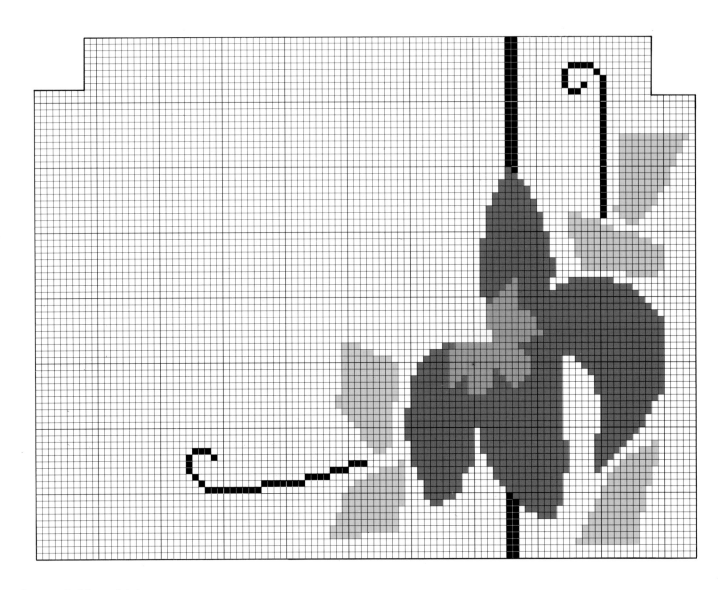

Incorporate this graph into
the back of the jacket,
remembering to continue the
vertical black stripe to the
top.

next row. Cont straight until 120 rows have been worked (ending with a WS row).
Shape neck: cast off 7 sts at beg of next row, 4 sts at beg of the following alt row, 3 sts beg the next alt row, then dec 1 st at neck edge on every row 6 times. Work 2 rows.
Shape shoulder: working straight at neck edge, cast off 14 sts at beg of next row, work 1 row, cast off remaining 13 sts.

Left front
Work as for right front reversing shapings.

Back
Using 3mm needles and red, cast on 105 sts. Row 1: k1, p1 to last st, k1. Row 2: p1, k1 to last st, p1. Rep these 2 rows twice more. Change to 3¾mm needles and blue. Commence following the graph in st st starting with a knit row until you have worked 70 rows. Cast off 7 sts at beg of next 2 rows.
Complete the graph, then work straight in blue, but cont the vertical stripe in black to the top of the back. Cont straight until the back matches the front in length to shoulder shaping (omitting the neck shaping). Cast off 14 sts at beg of next 4 rows. Cast off the remaining 35 sts.

Sleeves
Using 3mm needles and red, cast on 56 sts. Work in k1, p1 rib for 4 rows. Change to 3¾mm needles and blue. Commence following graph in st st starting with a knit row and increasing 1 st each end of every 3rd row until you have 106 sts. Work straight until graph is complete. Cast off loosely.

Buttonband
Using 3mm needles and red, cast on 7 sts. Row 1: k1, p1 to last st, k1. Row 2: p1, k1 to last st, p1. Rep these 2 rows until you have worked 130 rows or until buttonband reaches beg of the neck shaping when slightly stretched. Cast off and stitch band carefully to left front using a flat seam.

Buttonhole band
Using 3mm needles and red, cast on 7 sts. Rib 3 rows. Next row: p1, k1, cast off 3 sts, k1, p1. Return row: k1, p1, cast on 3 sts, p1, k1. Cont as for buttonband, making buttonholes as described, beg on rows 22, 38, 54, 72, 90, 108 and 126. Work 3 rows, cast off. Join band carefully to right front.

Neckband
Join shoulder seams using a flat seam and taking care to match the black vertical stripe. Using 3mm needles and red, pick up 22 sts up front of right neck (starting half way along the buttonband), 44 sts across back of neck, 22 sts down front of neck (88 sts). Work in k1, p1 rib for 5 rows. Cast off in rib.

Making up
Pin and sew sleeves into position. Join all seams using a flat seam (*see* Techniques, pages 18–19). Sew on buttons. Press lightly.

TUBE CAMISOLE

Using 4mm needles and blue, cast on 150 sts. Work straight in k1, p1 rib until work measures 27cm finishing with an RS row. Change to red, rib for another 6cm. Cast off. Join side edges using a flat seam in appropriate colours (*see* Techniques, pages 18–19).

BIG BOW SWEATER

A short-sleeved silky mercerized cotton sweater featuring a big bow and a mock sailor collar, worked using the intarsia method.

Materials
Melinda Coss 6-ply mercerized cotton – main colour: 450gm; contrast: 100gm.

Needles
One pair of 2¾mm and one pair of 3¾mm needles, one 2¾mm circular needle.

Tension
Using 3¾mm needles and measured over st st, 24 sts and 29 rows = 10cm square.

Back
Using 2¾mm needles and main colour, cast on 132 sts. K2, p2 rib for 22 rows. Change to 3¾mm needles* and, working in st st, work 4 rows in main colour, 8 rows contrast, 86 rows main. Commence following graph for back, changing colours as indicated until it is complete (row 172). Cast off loosely.

Front
Work as for back to *. St st 4 rows in main colour, 8 rows in contrast and 38 rows in main. Commence reading graph for front, changing colours accordingly until row 150.
Shape neck: pattern across 54 sts, cast off 24 sts for bottom neck edge and slip 54 sts, just worked on to a spare needle. Work to end. *Dec 1 st at neck edge on the next 5 rows, then dec 1 st at neck edge on the next row and every alt row until 44 sts remain. Cast off loosely. Repeat from * for other side.

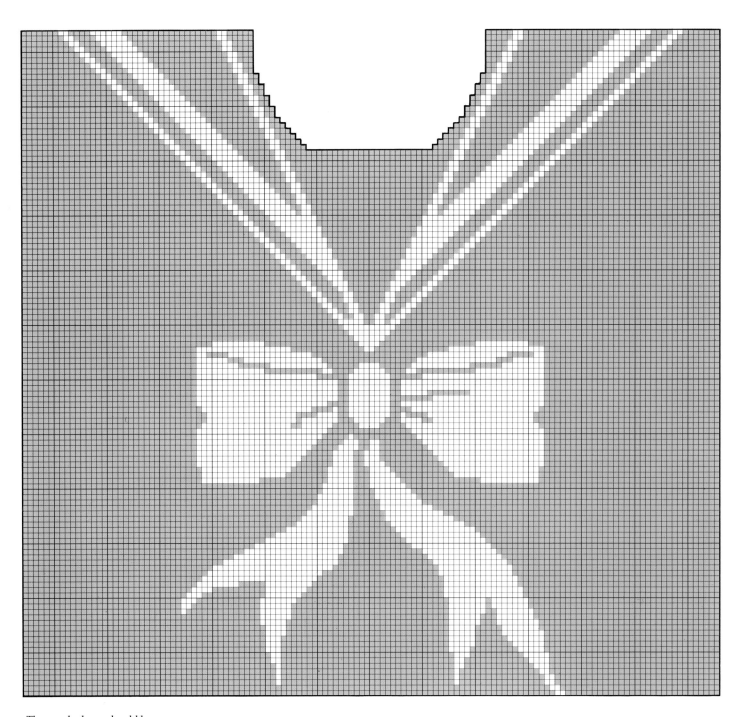

The graph above should be
incorporated into the front of
the sweater.

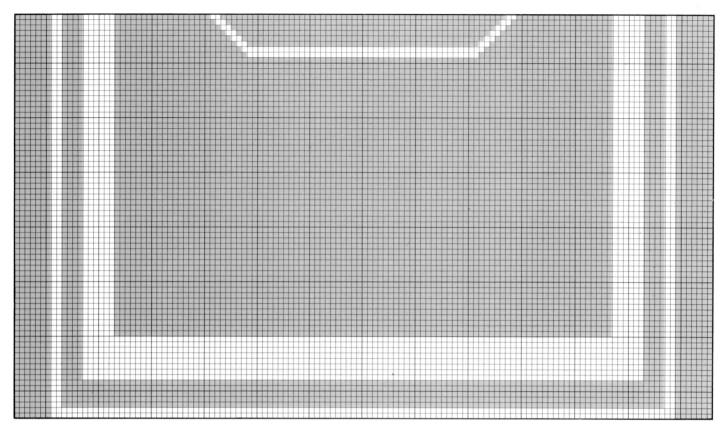

Sleeves

Using 2¾mm needles and main colour, cast on 88 sts. Work in k2, p2 rib for 14 rows. Change to 3¾mm needles and st st 4 rows in main, 8 rows in contrast and 20 rows in main. *At the same time*, commencing on row 3 of the st st, inc 1 st each end of this row and every alt row until you have 118 sts. Cast off loosely.

Neck edge

Using a 2¾mm circular needle and main colour, start at left shoulder seam and pick up 30 sts down left front neck edge to centre front, 30 sts up right neck edge and 40 sts across back neck edge. Work 8 rounds in k2, p2 rib. Cast off in rib.
Sew in all ends neatly.

Making up

Join shoulders together, matching collar motif, with a flat seam (*see* Techniques, pages 18–19). Mark depth of armholes 27cm down from shoulder seams and sew sleeves to body of sweater. Join underarm and side seams.

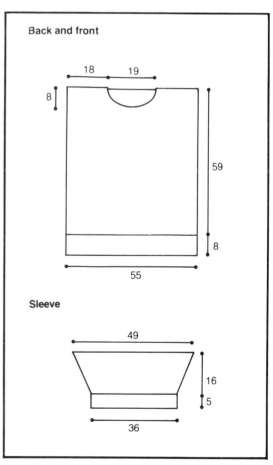

ADMIRAL'S JACKET

A sporty, double-breasted jacket worked in double-knit cotton, using the intarsia method.

Materials
Melinda Coss DK cotton – ecru: 800gm; navy: 100gm; 6 brass buttons.

Needles
One pair of 3mm and one pair of 4mm needles; one 3mm circular needle.

Tension
Using 4mm needles and measured over st st, 22 sts and 28 rows = 10cm square.

Pocket lining
Using 4mm needles and ecru, cast on 22 sts. Work 25 rows in st st and leave these sts on a spare needle.

Back
Using 3mm needles, cast on 100 sts. K2, p2 rib for 39 rows. Next row: rib 27, inc 1. *Rib 2, inc 1, rep from * 22 times, rib 27 (124 sts).. Change to 4mm needles. Cont in st st, using separate balls of contrasting yarn for each section of motif and positioning the graph on page 78 as follows: k1, work first row of graph, k1. Complete the 100 rows of the graph, then work 33 rows in garter st (knit every row). Cast off loosely.

Right front
Using 3mm needles, cast on 32 sts. K2, p2 rib for 39 rows. Next row: rib 5, inc 1. *Rib 2, inc 1; rep from * 10 times, rib 5 (44 sts). Change to 4mm needles and work in st st for 100 rows. Work the next 33 rows in garter st. Cast off loosely.

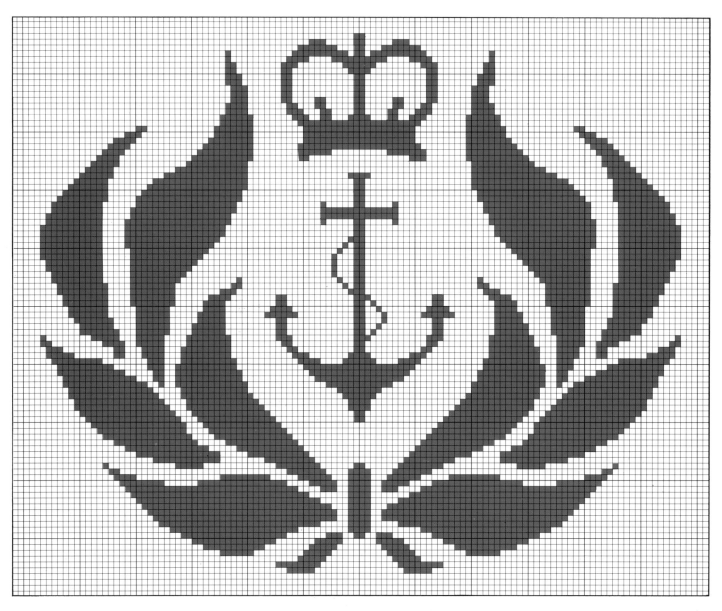

This graph should be
incorporated into the back of
the jacket.

Incorporate the graph (right) into the sleeves.

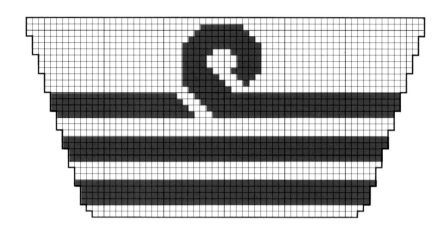

Left front

Work as for right front until 43 rows of st st have been worked. Next row: work across 11 sts. Work first row of the graph at the foot of the page. Work 11 sts to end. Cont working graph in this position until it is complete, purl across next row.

Make pocket: work across 11 sts, sl next 22 sts on to a stitch holder and work across sts of pocket lining. Work across 11 sts to end. Cont in st st as before until you have completed row 100. Work in garter st for 33 rows. Cast off loosely.

Sleeves

**Using 3mm needles and main colour, cast on 36 sts and work in k2, p2 rib for 24 rows. Next row: rib 11, inc 1. *Rib 2, inc 1; rep from * 6 times, rib 11 (44 sts remain). Change to 4mm needles.

Commence following sleeve graph above in st st. Inc 1 st each end of 1st and 3rd rows and then every 3rd row until you have completed the graph. Cont to inc on every 3rd row, work straight in st st until you have completed 116 rows (122 sts). Cast off loosely. Repeat from ** for second sleeve.

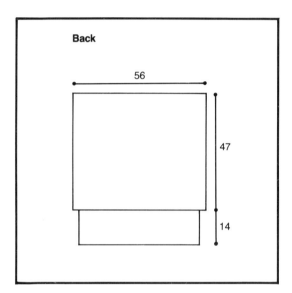

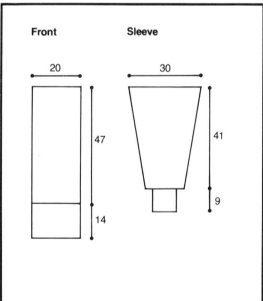

This graph should be incorporated into the left front.

Pocket opening

Using 4mm needles and main colour, pick up the sts held for pocket top and work 4 rows in k2, p2 rib. Cast off in rib.

Making up

Join shoulder seams, mark depth of armhole 28cm from shoulder seam, sew sleeves to body of sweater using a narrow backstitch, then sew side and underarm seams.

Shawl collar

Button side: using a 3mm circular needle, start at bottom, left front edge and pick up every stitch up front neck edge and across to centre back. Work in k2, p2 rib for 52 rows, turning work at end of each row and working back over stitches just worked. Cast off loosely in rib.

Buttonhole side: using a 3mm circular needle, start at bottom of right front edge and pick up every stitch up front edge and across to centre back. Work 3 rows in k2, p2 rib, turning work as before.

Row 4: rib 3 sts, make buttonhole (by casting off 2 sts, working to end and casting on 2 sts over those cast off on next row), rib 33 sts. Make buttonhole, rib 33 sts, make buttonhole, rib to end. Cont in rib until row 47. Make buttonholes as for row 4 on next row. Cont in rib until row 52. Cast off loosely in rib.

Place right front collar over left front collar and hem and mark positions for 6 buttons to correspond with buttonholes. Join back collar seam. Sew buttons into position. Sew in all ends. Press lightly using a damp cloth.

POLO SHIRT

A classic sports garment worked in 4-ply mercerized cotton in three sizes for men or women (*see* diagram for actual measurements). The collar, pocket top and so on are worked in garter stitch with a small contrasting stripe. **N.B.** If men's buttoning is required, reverse out the placket and buttonhole instructions.

Materials
Melinda Coss 4-ply mercerized cotton – navy: 350/400/450gm; brown: 50gm; 3 buttons.

Needles
One pair of 2¾mm and one pair of 3¼mm needles.

Tension
Using 3¼mm needles and measured over st st, 30 sts = 10cm.

Back
Using 2¾mm needles and navy, cast on 150/158/168 sts and work in garter st (knit every row) for 2cm. Change to 3¼mm needles and cont in st st until work measures 42/43/44cm.

Shape armholes: cast off 4 sts at beg of next 2 rows. Now dec 1 st each end of every row for 6 rows. Now dec 1 st each end of every alt row until 114/122/132 sts remain. Work straight until back measures 62/64/66cm, ending with a WS row.

Shape neck: k 36/39/43 sts, cast off 42/44/46 sts, k to end. Cont with this set of sts, leaving the others on a holder.

Next row: p to last 3 sts, p2 tog, p1.

Rows 2 and 3: k to last 11/12/13 sts, turn work, p to last 3 sts, p2 tog, p1.

Rows 4 and 5: k to last 22/24/26 sts, turn work, p to end. Leave sts on a holder. Return to other sts, joining yarn in at armhole edge.

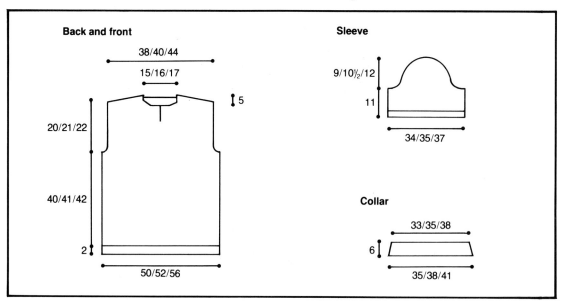

Back and front

38/40/44
15/16/17
5
20/21/22
40/41/42
2
50/52/56

Sleeve

9/10½/12
11
34/35/37

Collar

33/35/38
6
35/38/41

Next row: k to last 3 sts, k2 tog, k1.
Rows 2 and 3: p to last 11/12/13 sts, turn work, k to last 3 sts, k2 tog, k1.
Rows 4 and 5: p to last 22/24/26 sts, turn work and p to end. Leave sts on a spare needle/holder.

Front

As for back until work measures 42/43/44cm, ending with an RS row.

Shape armholes and divide for opening: next row: cast off 4 sts, p67/71/76 sts, k8 and turn work, leaving the remaining sts on a holder.

Row 2: k to last 3 sts, k2 tog, k1. Row 3: p1, p2 tog, p to last 8 sts, k8. Rep these last 2 rows twice more (69/73/78 sts).

Cont dec 1 st at armhole edge on next and every following alt row, maintaining the garter st border throughout, until 61/65/70 sts remain. Now work straight and on next RS row **work a buttonhole**: k3, cast off 2, k to end. On the following row, cast on 2 sts directly above those cast off on previous row. Cont working straight, making 2 more buttonholes, 5cm and then 10cm from the first (3 in all). Cont until placket measures 17cm from beg, ending with a WS row.

Shape neck: cast off 16 sts at beg of next row. Now dec 1 st at neck edge on every row until 34/37/41 sts remain. Work straight until work measures 62½/64½/66½cm, ending with a WS row. Next row: k to last 11/12/13 sts, turn work, p to end.

Next row: k to last 22/24/26 sts, turn and p to end. Leave these sts on a holder.

Using 3¼mm needles and navy, cast on 8 sts to form the button stand and join these in at the point where the sts were divided for the opening, purling to the end of the sts previously held. Next row: cast off 4 sts, k to end.

Work an 8 st garter st button stand, as for other side, but omitting the buttonholes.

Shape armhole: dec 1 st on every row until there are 69/73/78 sts and then every alt row until there are 61/65/70 sts. Now complete as for other side of neck, reversing out shapings. Leave sts on a holder.

Sleeve

Using 2¾mm needles and brown, cast on 99/103/107 sts, using the method described on pages 14–15. Change to navy and knit 2 rows. Next row: change to 3¼mm needles and carry the colour not in use, loosely, at back of work. K2 navy, *k3 brown, k1 navy, rep from * to last st, k1 navy. Row 2: purl in navy.

Return to 2¾mm needles and cont in navy in garter st for 6 rows.** Change to 3¼mm needles and cont in st st, inc 3 sts evenly across the first row (102/106/110 sts). Work straight until the sleeve measures 11cm.

Shape sleeve head: cast off 4 sts at beg of next 2 rows. Now dec 1 st each end of every row until 48/40/32 sts remain. Now cast off 4 sts at beg of every row until 8 sts remain. Cast these off.

Pocket

Using 3¼mm needles, cast on 31 sts and work in st st for 9½cm, ending with an RS row. Change to 2¾mm needles and knit 6 rows. Change to 3¼mm needles. Next row: p2 navy, *p3 brown, p1 navy, rep from * to last st, p1 navy. Return to 2¾mm needles and knit 2 rows in navy. Change to brown and cast off knitwise.

Collar

Using 2¾mm needles and brown, cast on 107/115/123 sts as for sleeve and work as for sleeve to **. Next row: k2, k2 tog, k to last 4 sts, k2 tog, k2. Cont in garter st, rep the dec row every 6th row until 99/107/115 sts remain. Cont in garter st until collar measures 6cm. Cast off loosely, using a larger needle.

Making up

Knit both shoulder seams together. Slip st the buttonband base down under the buttonhole band, taking care that the sts do not show on RS of work. Attach buttons to correspond to the buttonholes. Place a pin at the centre point on the top edge of each front band and attach the collar from pin to pin. Do so by its cast-off edge and using a flat seam.

Place pocket on breast, positioning according to taste. Slip st down on RS of work (*see* Techniques, page 19).

Join sleeve and side seams with a narrow backstitch. Pin the sleeves into the armholes, distributing the fabric evenly. Sew with a backstitch.

SEASCAPE

A simple, sleeveless sun-top in double-knit cotton, worked with a relief design of seashells, seaweed and starfish.

Materials
Melinda Coss DK cotton – 350gm.

Needles
One pair of 4mm needles.

Tension
Using 4mm needles and measured over st st, 21 sts and 25 rows = 10cm square.

Front and back
Using 4mm needles, cast on 98 sts and work in k1, p1 rib for 5cm. Commence following the graph starting with a knit row. All "×"

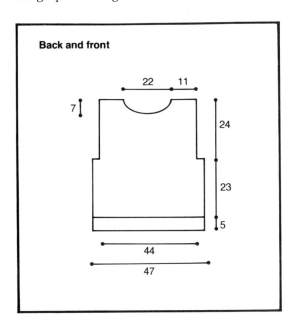

Back and front

symbols represent purl sts worked on the right side, otherwise st st throughout. The "seashells" are worked over 5 rows st st until you reach the 5th row of the symbol, then insert your right hand needle into the stitch marked ■ 5 rows down. *Draw through a loop, k1. Rep from * 6 times. On the return row, purl these 2 sts (the loop and the knit st) together 6 times. When 56 rows of the pattern have been worked, **shape armholes**: cast off 3 sts at beg of next 2 rows. Work without shaping until 100 rows of the pattern have been completed.
Shape neck (row 101): k37, turn, cast off 3 sts at beg of next row and next alt row. Cast off 2 sts at beg of the next 2 alt rows and 1 st at beg of the next 4 alt rows. Work 2 rows without shaping. Cast off. Rejoin yarn to centre stitches. Cast off 18 sts across front neck, shape left neck to match right, reversing shapings.

Making up
Join shoulder and side seams with flat seam (*see* Techniques, pages 18–19).

Incorporate the graph overleaf (page 86) into the front and back of the sweater; all "x" symbols represent purl stitches.

SEASHELL LACE

A fine, 4-ply drop-sleeved lace jumper worked in a seashell pattern using shiny mercerized cotton. The deep armholes and wide neckline provide a flattering shape for any figure.

Materials
Melinda Coss 4-ply mercerized cotton – 500gm.

Needles
One pair of 2¾mm and one pair of 3mm needles; one 2¾mm circular needle.

Tension
Using 3mm needles and measured over st st, 28 sts and 39 rows = 10cm square.

Front
Using 2¾mm needles, cast on 124 sts. Work 32 rows in k1, p1 rib, inc 26 sts evenly across the last row of the rib (150 sts). Change to 3mm needles.

Pattern
Each individual shell motif is made up as a panel of 20 sts and 16 rows and is described below. 5 sts should be knitted in garter st at beg and end of every row, i.e., k5, commence panel, rep panel to last 5 sts, k5. These 5 sts at beg and end of each row are *not* included in the panel instructions, so remember to work them.
Rows 1, 2 and 3: knit.
Row 4 (RS): *k10, (yo) twice, k10; rep from *.
Row 5: *k3, p7, make 5 new sts out of the double yo loop by working k1, p1, k1, p1, k1 into this loop; p7, k3; rep from *.
Row 6: *k2, ssk, k17, k2 tog, k2; rep from *.

Row 7: *k3, p17, k3; rep from *.
Row 8: *k2, ssk, k5, (yo k1) 5 times, yo, k5, k2 tog, k2; rep from *.
Row 9: *k3, p5, k11, p5, k3; rep from *.
Row 10: *k2, ssk, k19, k2 tog, k2; rep from *.
Row 11: *k3, p4, k11, p4, k3; rep from *.
Row 12: *k2, ssk, k2, (ssk, yo) 3 times, k1, (yo, k2 tog) 3 times, k2, k2 tog, k2; rep from *.
Row 13: *k3, p3, k11, p3, k3; rep from *.
Row 14: *k2, ssk, k15, k2 tog, k2; rep from *.
Row 15: *k3, p2, k11, p2, k3; rep from *.
Row 16: *k2 (ssk) twice, (yo, ssk) twice, yo, k1, (yo, k2 tog) 3 times, k4; rep from *.
Rep these 16 rows 12 times more, then cont in garter st only. Work 2 rows in garter st.
Shape neck: k58, cast off centre 34 sts, k58. Working each side of the neck separately, cast off 9 sts at beg of next 2 alt rows, then dec 1 st at neck edge on next 11 rows. Leave the 2 sets of remaining 29 sts for shoulders on spare needles.

Back
Work as for front.

Sleeves
Using 2¾mm needles, cast on 58 sts. Work in k1, p1 rib for 7cm, inc 28 sts evenly across last row of rib (86 sts). Change to 3mm needles and commence following the pattern as given for front and back but working 3 sts instead of 5 sts in garter st at beg and end of each row. Cont working in pattern until 10 complete sequences have been completed. *At the same time*, **shape sleeves**: inc 1 st each end of every 15th row until you have 106 sts. Complete shell pattern, then work 7 rows in garter st. Cast off very loosely.

Neckband
Placing front and back RS facing, use a 2¾mm needle and knit the shoulder seams together.
Using a 2¾mm circular needle, pick up and knit 160 sts evenly around the neck. K1, p1 rib for 6 rows. Cast off.

Making up
Lay the work out flat and pin sleeves into position, stitch sleeves to body using a flat seam. With work RS together and commencing at sleeve cuffs join sleeve and side seams.

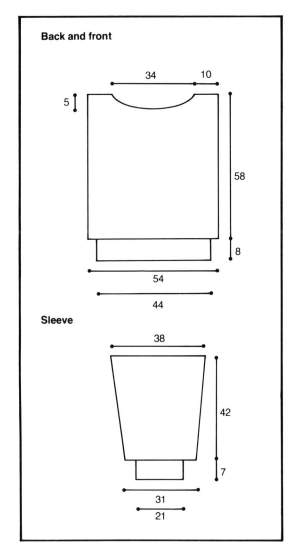

90

SLASH-NECK COTTON RELIEF JUMPER

A crisp white, double-knit cotton jumper with long sleeves and a wide neckline. Simple to knit, this pattern is ideal for beginners who would like to tackle a textured effect.

Materials
Melinda Coss DK cotton – 650gm.

Needles
One pair of 3¼mm and one pair of 4½mm needles.

Tension
Using 4½mm needles and measured over st st, 19 sts and 24 rows = 10cm square.

N.B. When reading the graph, work all sts with the symbol "○" in purl and all sts with the symbol "×" in knit. Where a small black square is shown, make a bobble – i.e., k3 from 1, turn, p3, turn, slip 1, k2 tog, psso. All sections of the graph without symbols should be worked in st st.

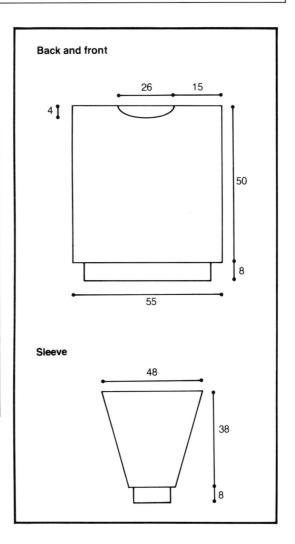

Front
Using 3¼mm needles, cast on 70 sts. K1, p1 rib for 20 rows, inc 34 sts evenly across last row of rib. Change to 4½mm needles and commence following the graph starting with a knit row. When you have completed the graph rep it once more, then rep it again until you have completed the first row of bobbles. Work 4 rows in st st, cont in st st only.
Shape neck: k36 sts, cast off centre 32 sts, k36 sts. Working each side of the neck separately, dec 1 st at neck edge on the next 8 rows. Cast off remaining 28 sts each side of neck.

Back
Work as for front to neck shaping. Cont straight to row 119. Cast off 28 sts at beg of next 2 rows, cast off remaining 48 sts.

Incorporate the graph below
into the front and back of the
sweater, repeating it as
instructed.

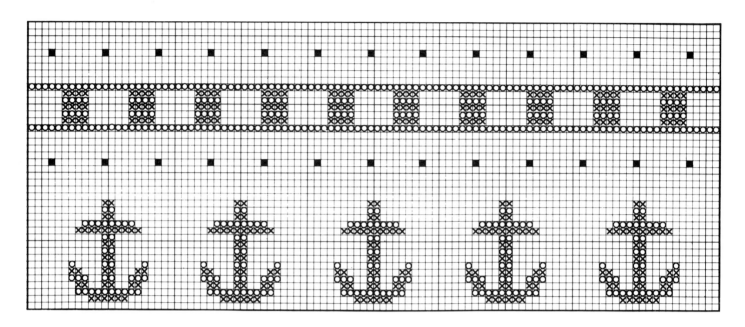

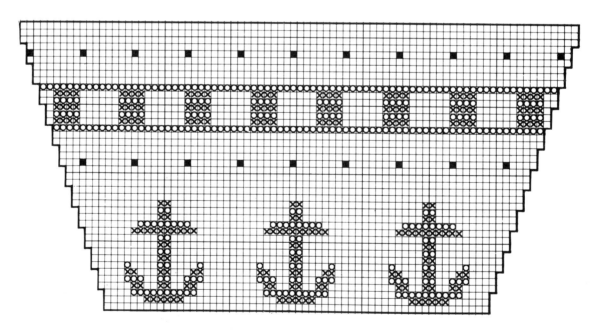

The graph above should be
incorporated into the sleeves,
repeating it and increasing
every third row as instructed
on page 94.

92

Sleeves

Using 3¼mm needles, cast on 38 sts. Work in k1, p1 rib for 20 rows, inc 20 sts evenly across last row of rib (58 sts). Change to 4½mm needles and commence following the graph as sectioned for sleeve, increasing 1 st each end of every 3rd row. Rep the graph once more, still increasing every 3rd row until you have 92 sts and then working without further shaping. When the 2nd repeat of the graph is complete, work 10 rows in st st, then cast off loosely.

Join one shoulder seam.

Neckband

Using 3¼mm needles, pick up 48 sts across back of neck, 6 sts down side of neck, 32 sts across centre front and 6 sts up side of neck. K1, p1 rib for 4 rows. Cast off in rib.

Making up

Join other shoulder seam and neckband, join sleeves to body and sew sleeve and body seams. Use flat seams throughout (*see* Techniques, pages 18–19).

TILTING LADDERS

An elegant, long-line lace jumper in a silky mercerized cotton – ideal for evening wear.

Materials
Melinda Coss 6-ply mercerized cotton – 700gm.

Needles
One pair of 3¼mm and one pair of 4mm needles.

Tension
Using 4mm needles and measured over st st, 24 sts and 29 rows = 10cm square.

Broken rib pattern
Row 1 (WS): *k1, p1.*
Row 2: knit.

Tilting ladder pattern
Row 1 (WS): k2, *p5, k1, p5, k2; rep from *.
Row 2: p2, *k1, (yo, k2 tog) twice, p1, k5, p2; rep from *.
Rows 3 and 5: k2, *p4, k2, p5, k2; rep from *.
Row 4: p2, *k1, (yo, k2 tog) twice, p2, k4, p2; rep from *.
Row 6: p2, *k1, (yo, k2 tog) twice, p2, sl next 2 sts to cable needle, hold at back, k2, then k2 from cable needle, p2; repeat from *.
Rows 7, 8, 9, 10, 11 and 12: rep rows 3, 4, 5 and 6, then rows 3 and 4 again.
Row 13: rep row 1.
Row 14: p2, *k5, p1, (ssk, yo) twice, k1, p2; rep from *.
Rows 15 and 17: k2, *p5, k2, p4, k2; rep from *.
Row 16: p2, *k4, p2, (ssk, yo) twice, k1, p2; rep from *.
Row 18: p2, *sl next 2 sts to cable needle and

95

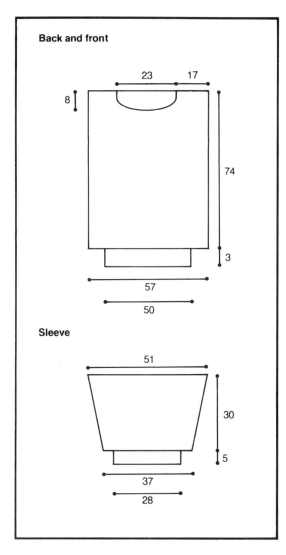

Back and front

23   17

8

74

3

57

50

Sleeve

51

30

5

37

28

Row 2: k22, p2. *K1, (yo, k2 tog) twice, p1, k5, p2; rep from * 6 times, k22.
Cont in pattern as set until work measures 74cm, ending with a WS row. Cast off 42 sts at beg of next 2 rows. Leave 53 sts on a stitch holder.

Front

Work as for back until front measures 66cm.
Shape neck (with RS facing): pattern 50 sts, turn and work on these sts. Pattern the next row and on the next row pattern to last 2 sts, k2 tog. Rep these last 2 rows until 42 sts remain. Cont until front length matches back, ending on a WS row. Cast off. Return to first set of sts, sl the first 37 sts on to a stitch holder, rejoin yarn and work left side to match right.

Sleeves

Using 3¼mm needles, cast on 70 sts and work in k1, p1 rib for 5cm, inc 20 sts evenly across the last row of rib (90 sts). Change to 4mm needles. Begin pattern:
(WS): k1, p1, k1, p1, k1 (broken rib pattern). K2, *p5, k1, p5, k2; rep from * 5 times. K1, p1, k1, p1, k1.
Next row: k5, p2, *k1, (yo, k2 tog) twice, p1, k5, p2; rep from * 5 times, k5.
Cont in pattern as set. *At the same time*, inc 1 st into the broken rib pattern at each end of next and every 4th row until you have 122 sts. Work straight until sleeve measures 35cm, ending with a WS row. Cast off loosely.

Neckband

Join right shoulder seam (*see* Techniques, pages 19–20). Using 3¼mm needles, pick up and knit 108 sts evenly around neck, including those sts left on st holders for back and front. K1, p1 rib for 3cm. Cast off.

Making up

Join left shoulder and neckband with flat seams (*see* Techniques, pages 18–19). Join sleeves to sides of back and front with shoulder seams in centre. Join side and sleeve seams.

hold in front, k2, then k2 from cable needle, p2, (ssk, yo) twice, k1, p2; rep from *.
Rows 19, 20, 21, 22, 23 and 24: rep rows 15, 16, 17 and 18, then rows 15 and 16 again.
Rep rows 1–24.

Back

Using 3¼mm needles, cast on 122 sts. K1, p1 rib for 9 rows, inc 15 sts evenly across last row of rib (137 sts). Change to 4mm needles and, following patterns as set out above, commence as follows:
Row 1 (WS): k1, p1 rib for 22 sts (broken rib pattern), k2. *P5, k1, p5, k2; rep from * 6 times. K1, p1 rib for next 22 sts.

CLASSIC CRICKET SWEATER WITH POCKETS

A long line double-knit cotton cricket sweater with a deep V-neck and boldly striped welts. Suitable for men and women.

Materials
Melinda Coss DK cotton – white: 900gm; navy: 50gm; yellow: 50gm.

Needles
One pair of 3mm and one pair of 4mm needles; one 3mm circular needle.

Tension
Using 4mm needles and measured over st st, 22 sts and 28 rows = 10cm square.

Abbreviations
c6b (cable 6 back): slip next 3 sts on to a cable needle and hold at the back of the work. K next 3 sts from left-hand needle, then knit sts from cable needle.

Pocket linings
(Make 2.) Using 4mm needles and white, cast on 28 sts. Work 28 rows in st st, leave sts on a spare needle.

Back
Using 3mm needles and white, cast on 122 sts and work 6 rows in k2, p2 rib. Change to navy, knit across 1 row and rib next row. Change to yellow, knit across 1 row and rib next 11 rows. Change to navy, knit across 1 row and rib next row. Change to white, knit across 1 row and rib next 5 rows.

Pattern
Row 1 (RS): *k6, p2; rep from * to end.
Row 2: k2, *p6, k2; rep from * to end.
Row 3: work as for row 1.
Row 4: work as for row 2.
Row 5: *p2, c6b, p2, k6; rep from * until last 10 sts. P2, c6b, p2.
Row 6: work as for row 2.

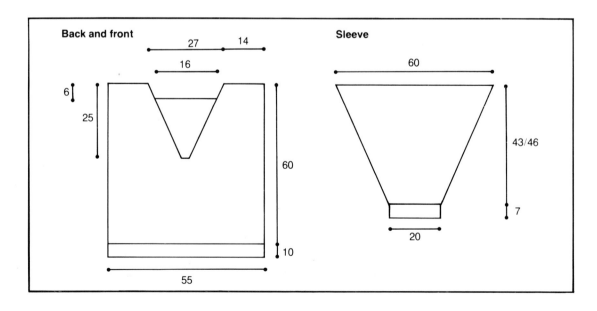

Rep 1st and 2nd rows twice more.
Rep the last 10 rows twice more, then rep 1st
and 2nd rows until another 20 rows have
been worked. These 50 rows form the
pattern. Work in pattern until row 150.
Shape neck: row 151: *pattern across 43 sts,
cast off 36 sts for bottom neck edge, slip 43
sts just worked on to a stitch holder. Pattern
to end.
Row 152: dec 1 st at neck edge on this and
every row until row 163.
Row 164: cont in pattern without further
shaping until row 168 (31 sts remain). Cast
off loosely. Rep from * for other side.

Front

Work as for back until row 28.
Row 29: make pockets: pattern across 17 sts,
sl next 28 sts on to a stitch holder and work
across stitches of one pocket lining, pattern
across 32 sts, sl next 28 sts on to a stitch
holder, work across stitches of 2nd pocket
lining, pattern to end. Cont in pattern until
row 98.
Shape neck: row 99: **pattern across 60 sts,
cast off 2 sts for bottom of "V", sl 60 sts just
worked on to a stitch holder. Pattern to end.
Row 100: dec 1 st at neck edge on this and
every row until row 112.
Row 113: dec 1 st at neck edge on this and
every other row until row 141.
Work a further 9 rows, dec 1 st on row 51.
Cont without further shaping at neck edge
until row 168 (31 sts remain). Cast off
loosely. Repeat from ** for other side.

Sleeves

Using 3mm needles and white, cast on 44 sts
and work 4 rows in k2, p2 rib. Change to
navy. Knit across 1 row and rib next row.
Change to yellow. Knit across 1 row and rib
next 7 rows. Change to navy. Knit across 1
row and rib next row. Change to white. Knit
across 1 row and rib next 3 rows.

Pattern

Row 1: k1. *P2, k3; rep from * until last 3 sts.
P2, k1.
Row 2: inc 1 st at both ends of this row and
the next 6 rows, cont to work in main pattern
as established. Work row 9 with no
increases.
Row 10: cont in main pattern, inc 1 st each
end of this and every 3rd row until you have
132 sts. *At the same time*, when you reach row
31 cont by working repeats of rows 1 and 2
only until row 120. (For man's size only,
work an extra 10 rows without shaping.)
Cast off loosely.

Neckband

Join shoulder seams with a flat seam (*see*
Techniques, pages 18–19). Using a 3mm
circular needle and white, begin at base of
the "V" and pick up 49 sts up right front, 58
sts across back of neck and 49 sts down left
front (156 sts). Work 20 rows in k2, p2 rib,
changing colours as per instructions for
sleeve ribs. Cast off in k2, p2 rib. With a
coloured thread mark 7½cm up from base of
the "V" on both sides. Place right end of
neck ribbing across to left marker and sew
neatly into position, then place left end
behind right end to right marker and sew in
place.

Pocket welts

Using 4mm needles and white, pick up
stitches held for pocket welts and work 6
rows in k2, p2 rib. Cast off loosely. Sew
pocket opening and lining to body of
sweater. Rep for second pocket. Sew in all
ends neatly.

Making up

Mark depth of armholes 30½cm down from
shoulder seams and sew sleeves to body of
sweater. Sew underarm and side seams
using a flat seam (*see* Techniques,
pages 18–19).

LOVE BIRDS

A cropped, boxy top in double-knit cotton, with a slash neck and dropped sleeves. Worked in stocking stitch using the intarsia method, it is accompanied by a very mini, mini skirt.

Materials
Melinda Coss DK cotton – **Jumper** white: 500gm; 50gm each of purple, lime, lemon, red, black, grey and coffee. **Skirt** white: 300gm.

Needles
One pair of 3mm and one pair of 3¾mm needles; one 3mm circular needle.

Tension
Using 3¾mm needles and measured over st st, 20 sts and 28 rows = 10cm square.

JUMPER

Front
Using 3mm needles, cast on 108 sts. K1, p1 rib for 10 rows. Change to 3¾mm needles and, starting with a knit row, commence following the graph in st st until you have worked 54 rows. Cast off 5 sts at beg of next 2 rows. Cont until you have completed 119 rows.

Shape neck: p25. Slip these sts on to a spare needle, cast off centre 48 sts, p to end.
Next row: cast off 12 sts, k8, k2 tog, turn, p2 tog, p to end.
Next row: cast off remaining 9 sts. Rejoin yarn to inner edge on second side and work to match.

102

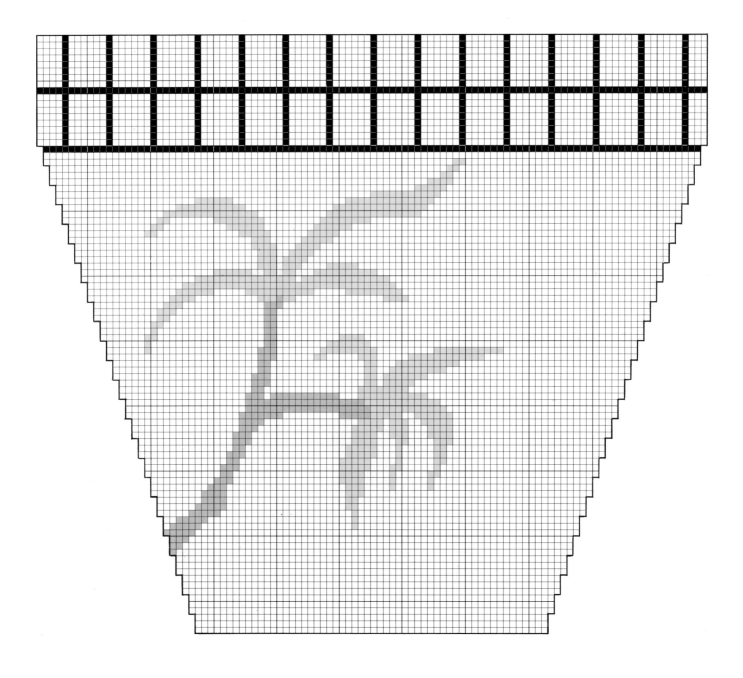

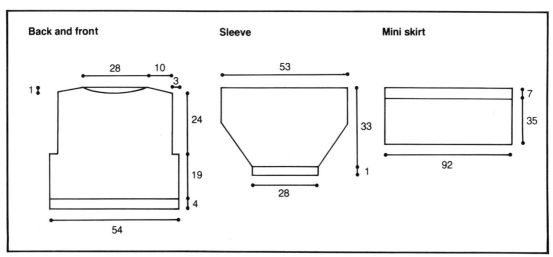

Back and front

28   10

3

1

24

19

54

4

Sleeve

53

33

1

28

Mini skirt

7

35

92

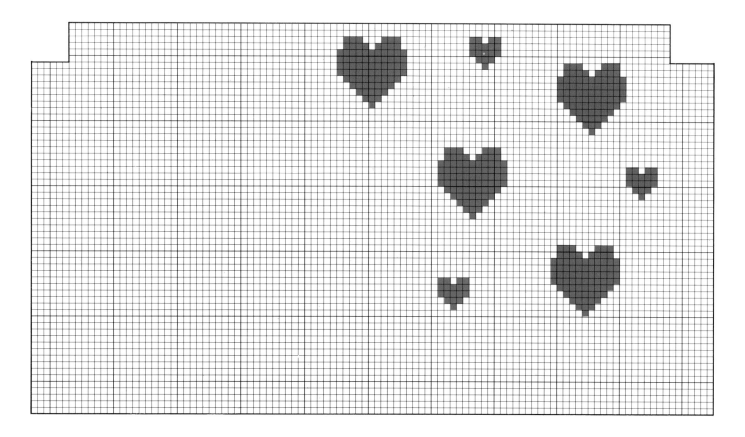

Back

Work as for front but following back graph. When graph is completed, cont in main colour only until length matches front to shoulder shaping. Shape shoulders and neck as for front. Cast off.

Sleeves

Using 3mm needles and white, cast on 56 sts. K1, p1 for 4 rows. Change to 3¾mm needles and, starting with a knit row, commence following graph in st st, inc 1 st each end of every 3rd row until you have 106 sts. Work without further shaping until graph is complete. Work 2 rows in black, cast off loosely. Join shoulder seams together using a flat seam.

Neckband

Using a 3mm circular needle, pick up and knit 56 sts evenly around the neck. K1, p1 rib for 4 rows. Cast off.

Making up

Backstitch sleeves to body, taking care to match the centre of the sleeve top with the shoulder seam. Join sleeve and side seams using a flat seam.

MINI SKIRT

Using 3¾mm needles, cast on 188 sts. Work in k2, p2 rib until work measures 35cm. Change to 3mm needles and work in st st for 7cm. Cast off.

Making up

Turn waistband inwards and slip st cast-off edge to beg of st st band. Invisibly join side seam of skirt, leaving an opening at the waistband for elastic. Cut a length of 2½cm elastic to fit comfortably around waist. Thread through waistband and sew ends together. Invisibly seam opening of waistband.

Incorporate the graph on page 102 into the front of the sweater. The graph opposite (page 104) should be incorporated into both sleeves, and the graph above into the back of the sweater.

SNOWFLAKE FAIRISLE

An oversized raglan sweater worked in one body width but two lengths (*see* diagram). This striking pattern is achieved using the fairisle technique of colour knitting (*see* Techniques, pages 12–13). Although worked in 4-ply mercerized cotton, the use of the two colours results in a double-thickness fabric.

Materials
Melinda Coss 4-ply mercerized cotton – natural: 500/550gm; black: 250/300gm.

Needles
One pair of 2¾mm and one pair of 3¼mm needles.

Tension
Using 3¼mm needles and measured over st st, 32 sts and 39 rows = 10cm square.

Back
Using 2¾mm needles and natural, cast on 200 sts.
Row 1 (WS): *k1, p1, rep from * to end. Row 2: *k1, tb1, p1, rep from * to end. Keep rep last 2 rows to form single twisted rib for 6/7cm, ending on a WS row. K next row, inc into every 10th st (220 sts).
Change to 3¼mm needles and start working graph from row 1/12. Work the sts of the graph 5 times across the row. Complete the graph rows. Work the 40-row pattern twice more and then work rows 1–11 inclusive.
Shape raglan (keeping in pattern throughout): cast off 8 sts at beg of next 2 rows.
Next row: work 2 tog, work to last 2 sts, work 2 tog. Row 2: rep first row. Row 3: work straight. Keep rep these last 3 rows until 72 sts remain. Now dec 1 st each end of every

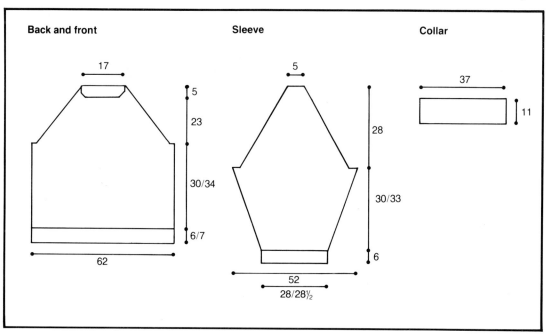

Back and front 17 / 5 / 23 / 30/34 / 6/7 / 62

Sleeve 5 / 28 / 30/33 / 6 / 52 / 28/28½

Collar 37 / 11

row until 6 patterns have been completed from beg (54 sts remain). Cast off.

Front

As for back until 90 sts remain.

Shape neck (meanwhile cont to shape raglan as before): work to central 20 sts, cast these off and work to end. Cont with this set of sts, leaving the others on a holder. Dec 1 st at neck edge on every row until 21 sts remain. Now dec 1 st at neck and raglan edge on every row until 7 sts remain. Now work this neck edge straight and cont to shape raglan on every row until 2 sts remain (and front matches back). Work 2 tog and pull yarn through to fasten off. Return to other side of neck and shape to match.

Sleeves

Using 2¾mm needles and natural, cast on 60/69 sts and work in single twisted rib as for back for 6cm, ending with an RS row. Purl the next row, inc into every alt/3rd st (90/92 sts). Change to 3¼mm needles and start working from graph from row 1/12. Inc 1 st at each end of the next 1/2 rows and every following 3rd row until there are 168 sts (working all new sts into pattern). Now work straight until the 11th row of the 4th pattern repeat has been worked.

Shape raglan (keeping in pattern throughout): cast off 4 sts at beg of next 2 rows. Dec 1 st each end of every alt row until 88 sts remain. Now dec 1 st each end of every row until 5 patterns have been worked from beg of sleeve (16 sts remaining). Cast off.

Collar

Using 2¾mm needles and main colour, cast on 171 sts and work in twisted single rib for 11cm. Cast off loosely in rib.

Making up

Join ribs with flat seams and use a narrow backstitch over pattern.

Sew the side edges of the collar with a flat seam for 3cm up from the cast-off edge. Place this partial seam at the exact centre of the front neck and attach with a flat seam.

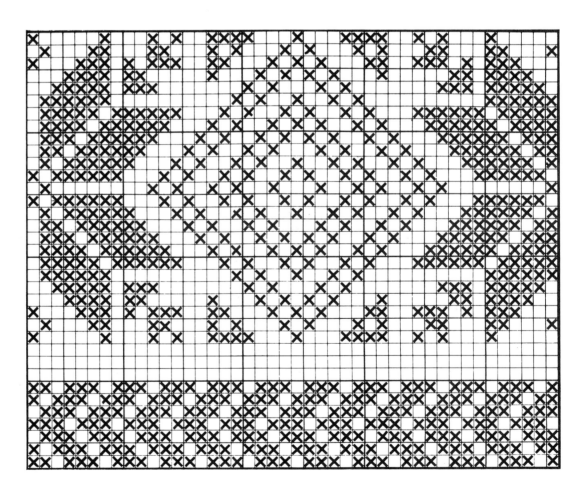

Incorporate this graph into the back, front and sleeves, repeating it as instructed.

TATTOO JACKET

A masculine shawl-necked, raglan jacket with amusing "tattoo" motifs on the back and sleeves. It is worked in a double-knit matt-finish cotton, and the instructions are given for men's medium/large.

Materials
Melinda Coss 8-ply matt cotton – natural: 1,100/1,150gm; 50gm each of black, red and green; one 60/65cm open-ended zip fastener.

Needles
One pair of 3¼mm and one pair of 3¾mm needles.

Tension
Using 3¾mm needles and measured over st st, 22 sts = 10cm.

Back
Using 3¼mm needles, cast on 126/134 sts.
Row 1: *k1, p1; rep from * to end. Keep rep this row to form single rib for 7cm, ending with a WS row. Knit the next row, inc into every 12th/13th st (136/144 sts).
Change to 3¾mm needles and cont in st st until work measures 29/30cm, ending with a WS row.
Next row: k45/49 in natural, work first row from main graph, k to end in natural. Keep working the graph in this position until it is complete. Meanwhile, when work measures 34/37cm, ending on a WS row, **shape raglan**: cast off 2 sts at beg of next two rows.
Row 3: k2, sl 1, k1, psso, k to last 4 sts, k2 tog, k2. Row 4: purl to end. Keep rep these last 2 rows until 72/80 sts remain. Now dec 1 st at each end of every row, in the same position as before, until 34 sts remain. Cast off.

Right front

Using 3¼mm needles, cast on 66/70 sts.
Row 1: rib to last 6 sts, k6. Row 2: k6, rib to
end. Cont thus, in single rib with a 6 st
garter st border, until work measures 7cm,
ending with a WS row. Knit the next row,
inc into every 12th st (71/75 sts). Change to
3¾mm needles and cont in st st, maintaining the garter st border the length of the
front. When work measures 14/15cm, ending
with an RS row, **work pocket slit**: k6, p39,
turn work, leaving remaining sts on a
holder. Cont with working sts until work
measures 28/29cm from beg. Now cast on 30
sts to form the pocket lining and join these
in at point where sts were split, working
across held sts (56/60 sts on the needle). Cont
with these sts until this section is as deep as
that previously worked, ending with an RS
row. Cast off the 30 lining sts, p to end. Next
row: knit across all sts. Now cont with the
full number of sts, as before, until the front
measures 34/37cm, ending with an RS row.
Shape raglan: cast off 2 sts at beg of next row.
Next row: k6, p to end. Row 2: k2, sl 1, k1,
psso, k to end. Keep rep these last 2 rows
until 39/43 sts remain. Now start dec 1 st at
raglan edge on every row. Meanwhile, when
work measures 60/65cm, ending with a WS
row, **shape neck**: cast off 9 sts at beg of next
row. Now dec 1 st at neck edge on every row
until 6 decs have been worked. Now work
this edge straight, meanwhile cont to dec
raglan until 2 sts remain. Work 2 tog and
fasten off.

Left front

Work as for right front, reversing out
instructions.

Sleeves

Using 3¼mm needles, cast on 50/54 sts and
work in single rib for 7cm, ending with a WS
row. Knit the next row, inc into every 5th st
(60/64 sts). Change to 3¾mm needles and
cont in st st, inc 1 st each end of every 4th
row until you have 114/118 sts, ending with
a WS row (cont working sleeve straight).
Next row: k46/48 in natural, work the first
row from the sleeve graph, k to end. Cont
working the graph in this position.
Meanwhile, when the sleeve measures
51/53cm, ending with a WS row, **shape
raglan**: cast off 2 sts at beg of next 2 rows.
Now shape raglan as for the back until 50 sts
remain. Now dec 1 st each end of every row
until 12 sts remain. Cast off.

Incorporate this graph into
the back of the jacket.

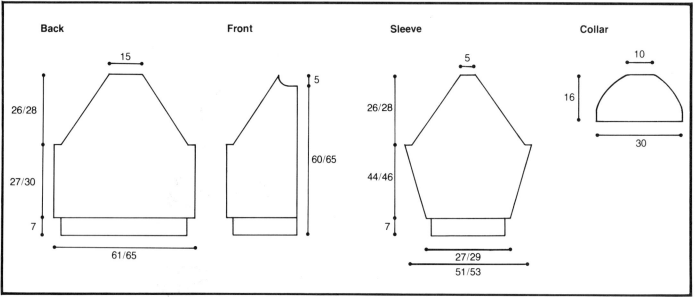

Back	Front	Sleeve	Collar

Back
15
26/28
27/30
7
61/65

Front
5
60/65

Sleeve
5
26/28
44/46
7
27/29
51/53

Collar
10
16
30

Collar

Using 3¾mm needles and natural, cast on
3 sts.
Row 1: k1, inc into next st, k1. Row 2: k1, inc
into next st, k to end. Cont in garter st, inc 1
st at the shaped edge on every row until you
have 23 sts and then every alt row until you
have 36 sts. Now work straight for 36 rows.
At the same edge, now dec 1 st on every alt
row until 23 sts remain and then on every
row until 3 sts remain. Cast off.

Pocket edges

Using 3¼mm needles, natural and with RS
of work facing, knit up 27/30 sts along the
pocket slit. Work in garter st until border is
2cm deep, ending with an RS row. Cast off,
knitwise.

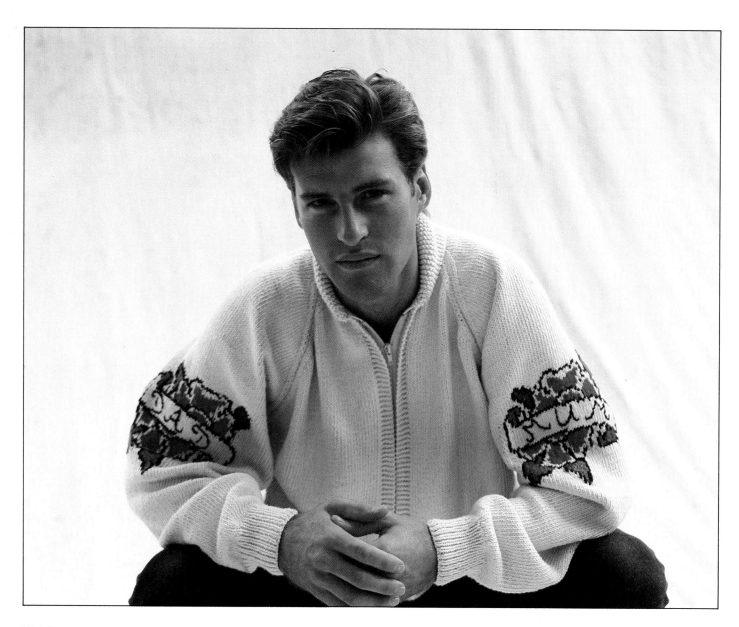

Making up

Join all seams with a flat seam (*see* Techniques, pages 18–19). Attach the shaped edge of the collar to the neck. Slip st down the pocket border edges and also the pocket linings, taking care that the stitches do not show on the RS of work.

Pin the zip fastener in position with the front edges of the knitting just touching and slip st down on the inside of the garter st border, using sewing thread rather than yarn.

Embroider names on the tattoo banners, according to taste (*see* Techniques, pages 20–21).

TATTOO VEST

A tongue-in-cheek macho singlet with tattoo motif. Worked in 6-ply matt cotton, sizes quoted are for women's/men's medium.

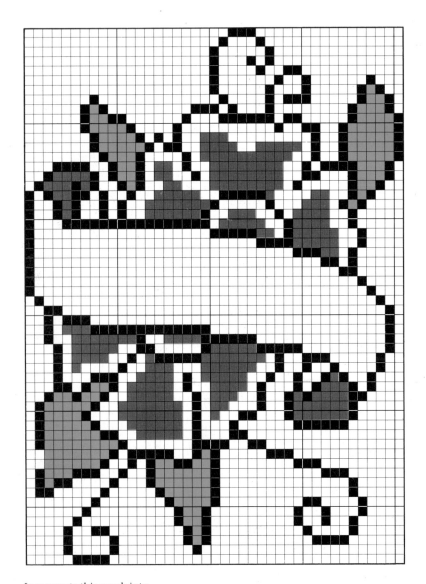

Incorporate this graph into the front of the vest and into the right and left sleeves of the jacket (see page 111).

Materials
Melinda Coss 6-ply matt cotton – natural: 250/300gm; 50gm each of black, red and green.

Needles
One pair of 3mm and one pair of 3¾mm needles.

Tension
Using 3¾mm needles and measured over st st, 26 sts = 10cm.

Front
Using 3mm needles and natural, cast on 102/116 sts.

Row 1: *k1, p1; rep from * to end. Keep rep this row to form single rib for 4cm. Change to 3¾mm needles and cont in st st, inc 1 st each end of the first row. When the work measures 20/22cm, ending with a WS row, **work the motif**: next row: k30/37, work the first row from the graph, k to end. Cont to work the graph in this position until it is complete. Meanwhile, when work measures 32/37cm, **shape armholes**: cast off 7 sts at beg of next 2 rows.

Next row: k2, k2 tog tb1, k to end, k2 tog, k2. Cont thus, dec 1 st each end of every row until 74/88 sts remain and then every alt row until 68/78 sts remain. Work 2/4 rows straight.

Shape neck: next row: work 22/25 sts, cast off 24/28 sts, work to end. Cont with this set of sts, leaving others on a holder. Now dec 1 st at neck edge on every row until 12 sts remain and then on every alt row until 8 sts

remain. Work straight until the front measures 57/64cm from beg. Leave sts on a holder. Return to other set of sts, join yarn in at neck edge and work to match first side, reversing out the shapings.

Back
As for front, omitting the motif, until work measures 32/37cm, ending with a WS row.
Shape armholes: cast off 7 sts at beg of next 2 rows.
Row 3: cast off 2 sts, k to last 4 sts, k2 tog, k2.
Row 4: cast off 2 sts, p to last 4 sts, p2 tog tb1, p2. Keep rep these 2 rows until 66/81 sts remain.
Dec 1 st at beg of next row and then dec 1 st each end of every row until 46/60 sts remain. Now dec 1 st each end of every alt row until 36/46 sts remain. Work straight until work measures 51/57cm.
Shape strap: inc 1 st at each end of next and every following 3rd/4th row. Next row: **shape neck** (meanwhile cont to shape strap): work 11/13 sts, cast off 14/20 sts, work to end. Cont with this set of sts, leaving the others on a holder. Dec 1 st at neck edge on every row until 8 sts remain. Now dec 1 st at neck edge on same rows as the incs on the outside edge of the strap until the back is as long as the front. Leave sts on a holder.

Return to other sts, join yarn in at neck edge and work to match first side.

Left armband
Knit the left shoulder seam tog. Open work out and using natural, 3mm needles and with RS facing, knit up 140/150 sts from front to back. Using a 3¾mm needle, cast off purlwise.

Right armband
Work as for left, but knit up from back to front.

Neckband
Front: using 3mm needles, natural and with RS facing, knit up 108/122 sts from the left shoulder seam to the right. Using a 3¾mm needle, cast off purlwise.
Back: work as for front, knitting up 40/48 sts from right to left shoulder and knitting up the first and last sts from the first and last sts of the front edging so that it becomes continuous.

Making up
Join side seams with a flat seam (*see* Techniques, pages 18–19).
Embroider on tattoo banner to taste (*see* Techniques, pages 20–21).

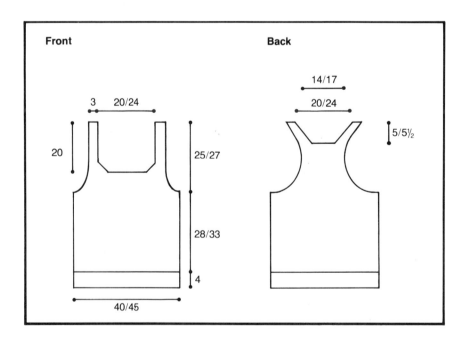

ME AND YOU KIDS' SWEATER

A fun crew-necked sweater for kids in double-knit cotton. Worked using the intarsia method, it is shown in two sizes to fit ages 4–5/6–8 years.

Materials
Melinda Coss DK cotton – main colour: 450/550gm; 50gm each of orange, yellow, apple, red and fuchsia.

Needles
One pair of 3mm and one pair of 3¾mm needles.

Tension
Using 3¾mm needles and measured over st st, 20 sts and 28 rows = 10cm square.

Front
Using 3mm needles, cast on 83/103 sts.
Row 1: k1, p1 to last st, k1.
Row 2: p1, k1 to last st, p1. Rep these 2 rows until rib measures 7cm. Change to 3¾mm needles and commence following the graph in st st until 99/105 rows have been worked.
Shape neck: pattern 31/39 sts. Cast off centre 21/25 sts. Slip first set of stitches on to a spare needle, work to end. Working on last set of stitches only, dec 2 sts at neck edge on next alt row, then dec 1 st at neck edge on next alt row. Work straight for 12/14 rows.
Shape shoulders: cast off 14/18 sts at beg of next row (shoulder edge). Work 1 row.
Cast off.

Back
Work as for front, omitting neck shaping.

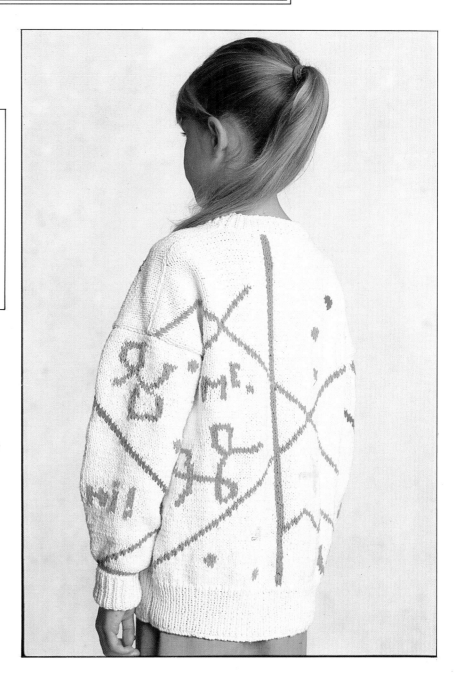

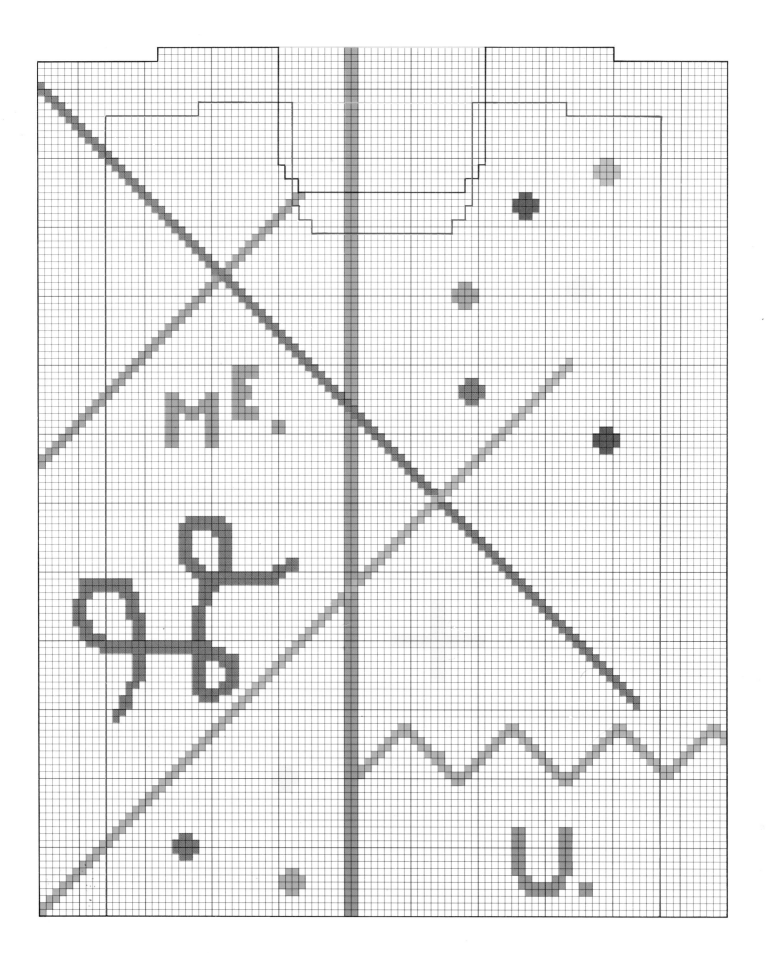

Incorporate the graph opposite (page 118) into the front and back of the sweater, following the red line for the smaller sized garment.

Sleeves

Using 3mm needles and main colour, cast on 36/38 sts and work 7cm in k1, p1 rib, inc 3/4 sts evenly across the last row of rib (39/42 sts). Commence following the graph working in st st and inc 1 st each end of every 5th row until you have 69/80 sts. Complete graph. Cast off loosely. Join one shoulder seam using a flat seam (*see* Techniques, pages 18–19).

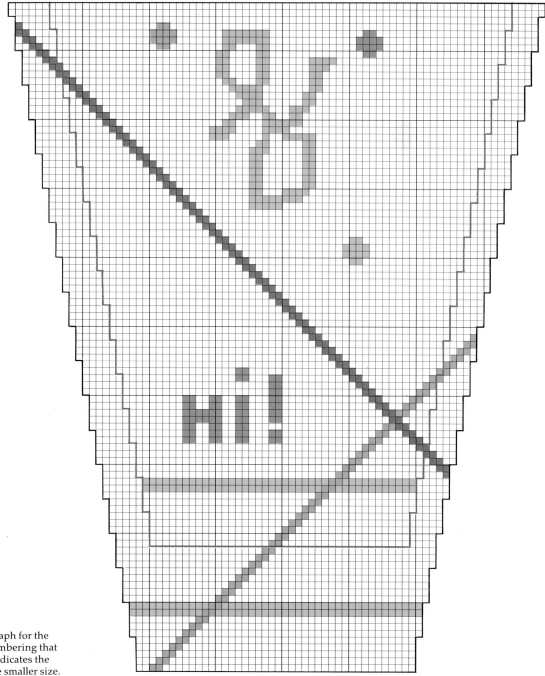

Follow this graph for the
sleeves, remembering that
the red line indicates the
pattern for the smaller size.

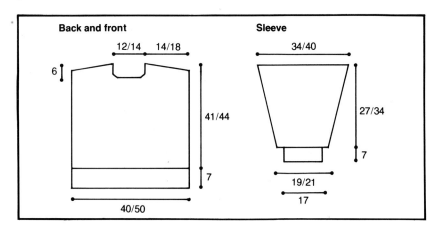

Back and front

12/14   14/18

6

41/44

7

40/50

Sleeve

34/40

27/34

7

19/21

17

Neckband

Using 3mm needles, pick up and knit 78/84
sts evenly around the neck. K1, p1 rib for
2cm. Cast off.

Making up

Join neckband using a flat seam. Pin and
sew sleeves to body and join sleeve and
body seams with a flat seam (*see* Techniques,
pages 18–19).

TODDLERS' DRESS/SWEATER

A bold baggy-fitting striped top for toddlers in sweater and dress lengths. Worked in matt 4-ply cotton , the sizes quoted are for 2–3/4–5-year-olds.

Materials
Melinda Coss 4-ply matt cotton – **Dress** (both sizes) white: 150gm; red: 150gm. **Sweater** white: 100/150gm; red: 100/150gm.

Needles
One pair each of 2¾mm, 3mm and 3¼mm needles.

Tension
Using 3¼mm needles and measured over st st, 28 sts and 34 rows = 10cm square.

DRESS

Front

Using 2¾mm needles and red, cast on 104/112 sts.

Row 1: *k1, p1; rep from * to end. Keep rep this row to form single rib for 2cm, ending with an RS row. Change to 3¼mm needles and cont in st st for 7 rows, inc 1 st each end of first row. Change to white and work 14 rows.Change to red and work 14 rows. Cont working in 14-row stripes until 9½/10½ stripes have been worked. Work 13 rows of next stripe. Next row: **shape neck**: k63/68 sts, put these sts on a holder and k to end across remaining sts. Cont with this set of sts, working in stripe pattern as before. Dec 1 st at neck edge on every row for 5 rows, then on every alt row until 33/36 sts remain. Now work straight until 1½ stripes have been worked from start of neck shaping. Leave stitches on a holder. Return to other

side of neck, leaving the centre 20/22 sts on a holder, shape to match first side. Leave stitches on a spare needle.

Back

Work as for front, omitting neck shaping.

Sleeves

Using 2¾mm needles and red, cast on 40/45 sts. Work in single rib for 4cm, ending with a WS row. Next row: knit, inc into every 5th st (48/54 sts). Change to 3¼mm needles and white and cont in stripe pattern, inc 1 st each end of every 3rd row until 4/4½ stripes have been worked, by which time there should be 84/96 sts on a needle. Cast off loosely.

Neckband

Knit the left shoulder seam tog (*see* Techniques, pages 19–20). Using 2¾mm needles, white and with RS of work facing, slip the 40/42 back neck sts on to a needle, then knit up 18 sts down the left side of neck, put held front neck sts on to another needle and knit across these, then knit up 18 sts up other side of neck (96/100 sts). Work in single rib for 1½cm. Cast off in rib.

Collar

Using 3mm needles and white, cast on 117/127 sts and work in single rib for 2 rows.

Change to red and knit the next row. Rib 1 row. Change to white, knit 1 row. Rib 1 row. Next row: k1, p1, sl 1, k2 tog, psso, cont in rib to last 5 sts, k3 tog, p1, k1.
Rib 3 more rows, then rep the dec row. Rep from * to *. Cont straight in rib until collar measures 5cm. Cast off.

SWEATER

Work as for dress, but having completed the welts, work a 14-row white stripe and cont in stripe pattern until last row of 8th/9th stripe before shaping the neck. Cont as for dress.

Making up

Knit left shoulder seam tog and join neckband with a flat seam (*see* Techniques, pages 18–19). Open out body and pin sleeves in position, taking great care not to bunch them (slightly stretch them to required armhole depth if in doubt). Join with a narrow backstitch. Join sleeve and side seams with a flat seam over ribs and a narrow backstitch over st st.
Pin collar in position, the cast-off edge to the knit-up line of neckband on inside of work (*see* photograph). Make sure that front edges are exactly centred at front of neck. Join with a flat seam.

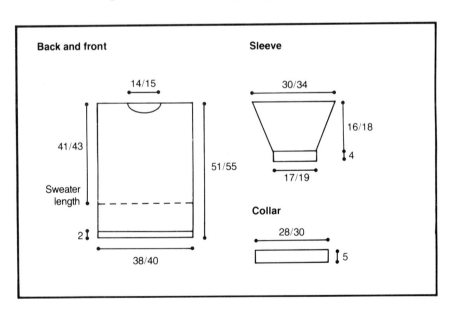

TRACKSUIT ROMPER

A sporty zippered romper for budding champions worked in a matt 4-ply cotton. Quoted in three sizes, for 9/12/18-month-old babies (see diagram for actual measurements).

Materials
Melinda Coss 4-ply matt cotton (all sizes) – red: 200gm; white: 50gm; 30cm zip fastener.

Needles
One pair each of 2¾mm, 3mm and 3¼mm needles.

Tension
Using 3¼mm needles and measured over st st, 28 sts = 10cm.

Back
Start at the lower edge of the left leg. Using 2¾mm needles and white, cast on 30/32/34 sts. Work in k1, p1 rib for 4 rows. Change to red and rib 2 rows. Rep last 6 rows, then rib 4 more rows in white. Knit the next row, inc into every alt st (45/48/51 sts). Now change to 3¼mm needles and red and cont in st st until work measures 18/19/20cm, ending with a WS row. **Shape crotch**: next row: cast off 3 sts, work to end.

Row 2: purl. Row 3: k1, k2 tog, k to end. Rep last 2 rows (40/43/46 sts). Leave these sts on a spare needle.

Work the right leg to match, reversing out the shapings and dec by k2 tog through backs of loops (tbl). Now slip the left leg sts on to the same needle, crotch shapings to the

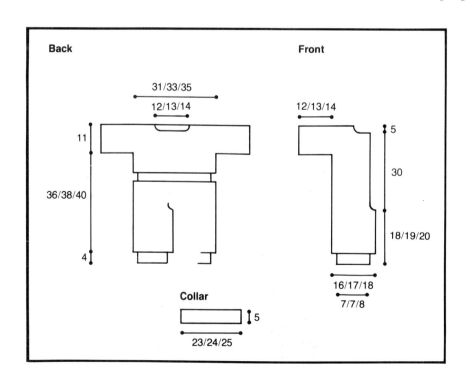

centre of the row.

Next row (WS): purl. Row 2: k37/40/43 sts, k2 tog tbl, k2, k2 tog, k to end (78/84/90 sts). Now inc 1 st each end of every 10th row until you have 88/94/100 sts. Work straight until work measures 34/36/38cm. Change to 2¾mm needles and white, and work in striped rib as at ankles, ending with 4 rows of white. Return to 3¼mm needles and red, and cont in st st until work measures 40/42/44cm. **Form sleeves**: cast on 33/36/39 sts at beg of next 2 rows (154/166/178 sts). Now work straight until work measures 51/53/55cm.

Next RS row: **shape neck**: k61/66/71 sts, cast off 32/34/36 sts, k to end. Cont with this set of sts, leaving others on a holder. Dec 1 st at neck edge on next 2 rows. Work 2 rows straight. Leave sts on a holder. Return to other side of neck and shape to match. Leave stitches on a holder.

Left front

Work from bottom of left leg as for back of right leg until work measures 18/19/20cm, ending with an RS row. **Shape crotch**: next row: cast off 4 sts, work to end.

Row 2: knit. Row 3: p1, p2 tog tbl, p to end. Rep these last 2 rows twice more (38/41/44 sts). Cont in st st but working a 3 st border in garter st (knit every row), in red, at front edge and inc 1 st at seam edge on every 10th row until you have 44/47/50 sts. Now cont as for back until work measures 40/42/44cm. **Form sleeve**: next RS row: cast on 33/36/39 sts, work to end. Cont straight until work measures 47/49/51cm. **Shape neck**: next WS row: cast off 8 sts and work to end. Dec 1 st

at neck edge on every row until 59/64/69 sts remain. Now work straight until front matches back. Leave sts on a holder.

Right front

This is worked as for left front, reversing out the shapings. When front and back have been worked, knit one shoulder seam tog (*see* Techniques, pages 19–20), cast off the back neck sts and knit tog the second shoulder seam.

Right cuff

Using 2¾mm needles, white and with RS facing, knit up 50/52/54 sts from back edge to front edge of cuff. Work a rib as for ankles, but ending with 3 rows in white. Cast off. Work left cuff the same, but knitting up sts from front to back edge.

Collar

Using 3mm needles and white, cast on 99/103/107 sts and work in stripe rib as at ankles but changing to 2¾mm needles after 12 rows and ending with 6 rows in white. Cast off in rib.

Making up

Set the zip fastener into the front opening so that the edges of the garter st borders just touch. Using sewing thread, slip st the zip into position, taking care that these sts are not visible on RS of work. Join sleeve, side and leg seams with a flat seam on ribs, a narrow backstitch over the st st.

Attach the collar to the neck by the cast-off edge. Pin from front edge to front edge and sew with a flat seam.

YARN INFORMATION

All the sample garments illustrated in this book were knitted in Melinda Coss yarns. As many of the designs contain small quantities of several different colours, Melinda Coss offers individual kits containing only the quantities of yarn necessary to complete each garment. In addition, buttons, embroidery threads and trimmings are included where appropriate. Contact Melinda at 1 Copenhagen Street, London N1 0JB (telephone: 01-833 3929).

For those who wish to substitute different yarns, weights are given throughout to the nearest 25gm ball. To obtain the best results you must ensure that the tension recommended on your selected yarn *matches the tension* printed in our pattern. We cannot guarantee your results if this rule is not followed.